DISTURBING THE SOLAR SYSTEM

DISTURBING THE SOLAR SYSTEM

Impacts, Close Encounters, and Coming Attractions

ALAN E. RUBIN

PRINCETON UNIVERSITY PRESS

PRINCETON AND OXFORD

Second printing, and first paperback printing, 2004
Paperback ISBN 0-691-11743-8

The Library of Congress has cataloged the cloth edition of this book as follows

Rubin, Alan E. (Alan Edward), 1953–
Disturbing the solar system : impacts, close encounters, and coming attractions /
Alan E. Rubin.
p. cm.
Includes bibliographical references and index.
ISBN 0-691-07474-7 (acid-free paper)
1. Gravity. 2. Catastrophes (Geology). 3. Life on other planets. I. Title.
QB339 .R84 2002
521'.1 — dc21   2001055197

British Library Cataloging-in-Publication Data is available

This book has been composed in Sabon with ITC Kabel Display

Printed on acid-free paper. ∞

pup.princeton.edu

Printed in the United States of America

To

Dorene,

David, Josh, and Jeremiah

Let us draw a lesson from nature, which always works by short ways. When the fruit is ripe, it falls. When the fruit is despatched, the leaf falls. The circuit of the waters is mere falling. The walking of man and all animals is a falling forward. All our manual labor and works of strength, as prying, splitting, digging, rowing, and so forth, are done by dint of continual falling, and the globe, earth, moon, comet, sun, star, fall for ever and ever.

—Ralph Waldo Emerson, *Spiritual Laws*, 1841

CONTENTS

○

PREFACE

O

I DOUBT IF AN apple ever fell on Isaac Newton's head, but there is no doubt that grand ideas raced through it. Newton's greatest achievement was to realize that the same force that pulls apples to the earth allows the Moon to orbit Earth and allows Jupiter's (then recently discovered) moons to orbit Jupiter. Newton called this force gravity and, in those pre–general-relativity days, imagined that it operated at a distance, in some spooky way between physically unconnected bodies. Although Newton's theory was wrong, his model is a good description of how most macroscopic objects behave when they are not close to very massive bodies.

It is gravity that was mainly responsible for the collapse of the cloud of gas and dust that formed our solar system. It is gravity that keeps the Sun from flying apart as it transforms hydrogen into helium in its core and generates tremendous outward pressure. It is gravity that binds asteroids and comets to the Sun, gravity that sometimes perturbs the orbits of these small objects and sends them careening around the solar system, and gravity that causes some to crash into a planet.

Gravity is what makes the Sun and its retinue of planets a *sys-*

tem, a word defined as a group of interacting and interdependent parts constituting a complex whole. After Newton published his theory of gravity in 1687, astronomers envisioned a "clockwork universe" analogous to a well-oiled machine in which gravity was responsible for keeping planetary bodies in their assigned orbits. But, as scientists began to realize in the ensuing centuries, the parts of the solar system are fundamentally interrelated; a disturbance in one locale can eventually produce dramatic consequences somewhere else. This book documents some of the profound indirect effects of gravity — the crash of a comet into Jupiter, the formation of planetary rings, the collisions of asteroids with Earth, the formation of the Moon, the death of the dinosaurs, the onset of extensive terrestrial glaciation, and the possible transfer of living organisms from one planet to another.

The latter topic takes us into the realm of astrobiology, a field sometimes unfairly disparaged as a discipline lacking a subject. But the purview of astrobiology is broad, encompassing the origin of life on Earth, the possibility that life could have evolved independently on Mars or Jupiter's moon Europa, or the chance that hypothetical alien civilizations are trying to dial us up right now.

To tell these tales of the solar system, I have looked through the literature on the history and philosophy of science, read the Bible, thumbed through numerous encyclopedias, examined astronomy, biology, geology, chemistry, and history textbooks, pored over scientific research and review papers (old and new), perused science magazines and trade science books, talked to experts in various fields, checked appropriate websites, read some poetry and classic science fiction, and even rented a few videos.

Many chapters are written with a historical perspective, and it is in this sense (as well as in the choice of topics) that this book differs from many others on the market. I have tried to show that science is an imperfect enterprise. I make my living as a research scientist and know how many mistakes my colleagues make. I am dismayed to realize that they sometimes say the same thing about their colleagues.

There is often great uncertainty at the frontiers of knowledge: some evidence seems to point in one direction, contradictory evidence in another. Competing scientists can seize upon different experimental or theoretical constraints and arrive at disparate conclusions. Impolite exchanges occur occasionally at scientific meetings from champions of opposing interpretations of available data. From time to time, students can be heard berating professors and professors belittling students. Scientists sometimes approach problems with preconceived notions of the "proper" interpretation of the evidence. Some researchers are reluctant to change their views even when faced with contradictory evidence that their colleagues find compelling. Experimental techniques can be flawed, and repetition of poorly designed experiments is not a sure path to truth.

Despite these problems (and they are not insignificant), the scientific endeavor is self-correcting. Claims of major discoveries are tested by other researchers in the field. These researchers may include skeptics searching for evidence contradicting the original claim, as well as excited proponents trying to ferret out additional aspects of the discovery that were initially overlooked. The motives of these researchers are irrelevant. In the long run, verified findings appear in review papers and textbooks; unverified claims such as the widely publicized "discoveries" of n-rays in the early twentieth century, polywater in mid-century, and cold fusion in 1989 fade into historical curiosities.

Eight of the chapters in this book are revisions of articles I wrote for the *Griffith Observer*, and I am particularly grateful to the editors of that publication for having provided a forum for my space-science vignettes over the years. Seven of those articles won awards in their annual science-writing contests.

The original version of chapter 3 appeared as "Heat Sources in the Solar System" in *Griffith Observer* 53, 2–11, 1989; chapter 4 as "The Magnetic Earth" in *Griffith Observer* 44, 2–9, 1980; chapter 6 as "Whence Came the Moon?" in *Sky & Telescope* 68, 389–93, 1984; portions of chapter 7 as "The Missing Planet" in

Griffith Observer 42, 10–17, 1978; chapter 9 as "A History of the Mesosiderite Asteroid" in *American Scientist* 85, 26–35, 1997; chapter 10 as "The Great Depression" in *Griffith Observer* 41, 2–8, 1977; a very small portion of chapter 12 as "Meteorites" in *Geotimes* 44, 55–56, 1999; chapter 14 as "Glass Menagerie" in *Griffith Observer* 43, 2–9, 1979; chapter 16 as "The Search for Life on Mars: Oasis or Mirage?" in *Griffith Observer* 62, 2–16, 1998; chapter 18 as "Paucity of Aliens" in *Griffith Observer* 61, 2–14, 1997; and chapter 19 as "The Human Response to First Contact" in *Griffith Observer* 64, 2–15, 2000.

I thank Ed Krupp of Griffith Observatory and the editors of *Geotimes, American Scientist,* and *Sky & Telescope* for permission to revise and republish these articles. I gratefully acknowledge the help I received from Joel Schiff of *Meteorite* in obtaining illustrations. Additional figures were provided by the following individuals: John Wasson, Frank Kyte, and Ned Wright of UCLA; Jeff Taylor of the University of Hawaii; Jim Sparling of Roosevelt University; Marty Prinz of the American Museum of Natural History; Ed Krupp of Griffith Observatory; Carleton Moore of Arizona State University; Richard Norton of Science Graphics; and Blaine Reed, a friendly meteorite dealer from Colorado. I thank them all. Cecelia Satterwhite and Marilyn Lindstrom of NASA-Johnson Space Center, Mary Noel and Paul Schenk of the Lunar and Planetary Institute, and Helen Worth of Johns Hopkins University obtained some NASA images. Dennis Case, archivist of the Air Force Historical Research Agency, provided the photograph of World War II bomb craters in Germany. I am also most grateful to Dorothy Norton of Science Graphics for her excellent line drawings.

I thank Jack Repcheck of W. W. Norton for looking favorably on the idea of a book and passing the manuscript along to Princeton University Press (PUP). Ed Scott of the University of Hawaii suggested several ways to make the manuscript more unified. Although these suggestions cost me a lot of time and involved a great deal of work, the book is better for them. I appreciate his advice and thank him for his efforts. I am grateful to the editors and staff at PUP for shepherding the manuscript to publication.

I also thank my family for putting up with my absences while I worked on the book. Even though I often wrote at home after everyone had gone to bed, I still had to steal away on the occasional weekend to my office at UCLA.

A final note to the reader about the designation of dates: Following modern practice, I use C.E. for common era instead of A.D., and B.C.E. (before the common era) instead of B.C.

Part 1

OVERVIEW OF THE SOLAR SYSTEM

WITH THE RECENT discovery of more than seventy extrasolar planets, it has become clear that there are many "solar" systems. They all likely formed in a manner rather similar to ours—from a rotating cloud of gas and dust that collapsed in on itself, rotated more rapidly, heated up, and eventually cooled down. Inside the cloud at various distances from the central protostar and at different times during the solar system's formation, matter was being processed and transformed in a variety of ways. An analytically inclined and long-lived eyewitness would have documented such varied processes as vaporization (matter heated until it becomes a gas), condensation (matter cooling from a gas to a condensed state), and accretion (matter colliding at low velocities and sticking together). Such a witness could observe matter interacting with the surrounding interstellar medium, propagation of shock waves through the cloud of dust and gas, irradiation of solid materials by the protostar, magnetic fields of varying intensities, gravitational perturbations, and numerous low-energy and high-energy collisions. In many stellar systems, and probably in our own, multiple stars formed simultaneously in small clusters or stellar associations.

In this section we will review the origin of the solar system and how astronomers deduced the location of the solar system in the Galaxy. As we will see later, this location is a favorable one that permits life to flourish. We will take a survey of available heat sources in the solar system to see how this energy affects the local landscape. We note that without heat, the solar system would be sterile, and without life, biologists would be out of work.

1

A BRIEF HISTORY OF
THE SOLAR SYSTEM

○

Some people feel . . . that by delving into the inner workings
of something, we destroy the essential beauty and mystery
surrounding it. From my point of view, nothing could be
farther from the truth. . . . A rainbow is just as beautiful to
someone who understands how it works as to anyone else."
—James Trefil, *Meditations at 10,000 Feet,* 1985

THE PREHISTORY of the solar system is an astronomical saga of star birth and death, of matter collapsing into gravitationally bound clouds of gas and dust and elements being
spewed into interstellar space. As the solar system formed, the
story shifts to one in which gravitational perturbations and collisions between multikilometer-size objects play dominant roles.

Stars spend most of their lives in a state of balance between
their tremendous gravity, which pushes matter inward, and the
gas pressure caused by the energy released during nuclear fusion —
the transformation of hydrogen into helium — which pushes matter
outward (figure 1.1). But toward the end of their lives, stars run

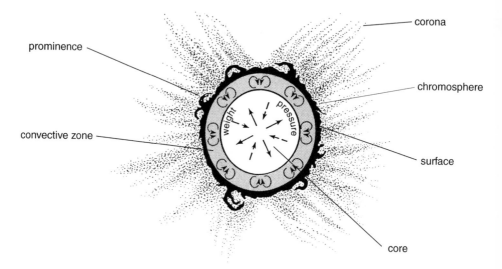

Figure 1.1. A stable star exists in a state of dynamic equilibrium in which the enormous weight of gas pushing inward is matched by the outward gas pressure caused by nuclear fusion in the stellar core. Surrounding the core is the convective zone, where hot parcels of gas move toward the surface, release their energy, cool down, and sink. The outermost atmospheric layer is the corona, which expands out from the star as a solar wind.

low on hydrogen and begin to release energy by creating heavier elements in their nuclear-fusion furnaces. Very massive stars can fuse helium into carbon and oxygen, and carbon into oxygen, neon, sodium and magnesium. Eventually silicon, phosphorus, and sulfur are produced and fused into iron. The fusion of heavy elements is a massive star's attempt to stave off the crushing effects of gravity, but it is a competition the star is doomed to lose. When enough iron is built up in the core, the energy spigot is turned off because it takes extra energy to split iron into lighter elements or fuse it into heavier ones.

Gravity then causes the core to collapse. As the core contracts, it heats up, eventually reaching temperatures of several billion degrees. Enormously energetic photons produced at these temperatures break down the iron nuclei into helium. This transformation uses up heat and cools the core. Gravity causes the core to collapse again. Lighter elements rain down on the core and fuse explosively into heavier ones. The massive star becomes a supernova as a shock wave blasts subatomic particles, heavy elements, and energetic photons out from the core and into the interstellar environment.

The newly formed heavy elements mix with ancient molecules that had previously escaped from the outer atmospheres of red giant stars and planetary nebulae. (These latter misnamed objects are expanding, nearly spherical clouds of hot gas ejected from old stars. They are not planets.) Much of this interstellar material becomes concentrated in giant molecular clouds such as the Orion Nebula, one of the most massive objects in the Galaxy. Giant molecular clouds are typically 300 light-years in diameter and range in mass from that of 100,000 to 1 million Suns. They are vast reservoirs of interstellar gas and dust held together by magnetic fields and turbulent motions. They contain numerous smaller dark clouds 5–15 light-years across. Gravitational instabilities, decreases in gas turbulence, leaking magnetic fields, and nearby supernova explosions can cause the small clouds to fragment, some into clumps only a few times as massive as the Sun. Loosely bound, multiple star systems known as stellar associations form from such clumps.

It seems likely that in some stellar associations, planets never form—mineral grains accreting around one star may evaporate after encountering sporadic intense radiation from a large, hot neighboring star.

The cloud that formed our solar system began to rotate more rapidly as it contracted. A significant fraction of the gas, dust, and ice was compressed into a disk that planetary scientists call the solar nebula. (The planets eventually formed from this disk; this is why most planets orbit the Sun in nearly the same plane.)

The Sun's equator is presently tilted about 6° relative to the mean orbital plane of the planets around the Sun. This tilt may reflect gravitational perturbation of the nebular disk by the close passage of a star. The star itself may have been a small errant sibling of the Sun that wandered off shortly after birth. Studies of star clusters have shown that small stars tend to escape from the clusters relatively quickly. This "evaporation" process eventually leads to disintegration of small clusters.

Interstellar gas and dust fell into the solar nebula, causing it to heat up. Material flowed into the early Sun, and it eventually became so massive that gravity exerted enough pressure in the solar interior to spark nuclear fusion. Magnetic forces caused jetting of molecular gas at the poles. The narrow gas jets, which may have extended for several light-years, carried angular momentum away from the Sun. Accretion of material continued. Eventually the supply of infalling material ran out, and the jets shut off.

Early in its history, the Sun was chaotic, flaring up sporadically and depositing large amounts of energy into the nebula. (Even at middle age, the Sun is somewhat variable, exhibiting a 22-year periodicity in solar activity. The Sun is 1 percent brighter about every 11 years, near the times when the maximum number of sunspots occurs.)

As the hot nebular gas around the nascent Sun cooled, silicate grains and other minerals condensed (i.e., came out of the gas phase as small solid particles). At lower temperatures, some fine-grained condensate particles aggregated into dust grains; other grains reacted with the gas to form new minerals.

The young energetic Sun generated powerful solar winds, driving water vapor and other gaseous materials out to about 5 AU, where the temperature was cold enough for water to condense into ice crystals. (An AU is an astronomical unit, equivalent to the mean distance between Earth and the Sun, about 150 million kilometers or 93 million miles.) The self-gravity of all this ice enabled an icy planet of about 10 Earth-masses to accrete quickly. This icy planet became the core of Jupiter; its large gravity field caused vast amounts of gas to be swept up from the nebula as it orbited the Sun.

At about the same time that Jupiter was forming, much else was going on in the solar nebula. Poorly understood accretionary processes were transforming dust into rocks and rocks into asteroid-size bodies known as planetesimals. Because meteorite researchers can study chips of these planetesimals and use them to infer the nature of the nebula, it is useful to recount what we think we know of how planetesimals formed.

High-energy events of some kind caused fine-grained nebular materials to evaporate, leaving behind refractory residues (materials that require high temperatures to melt or vaporize). Small chunks of rock in the primitive meteorites known as chondrites formed from these residues as well as from condensates. These chunks are rich in the refractory elements calcium, aluminum, and titanium and are known as refractory inclusions, calcium-aluminum inclusions, or CAIs. Some of the CAIs were remelted.

As the nebula cooled, less-refractory materials condensed and accreted. Porous clumps of silicate-rich dust were transformed into submillimeter-size spherules of melt, possibly after having been blasted by bolts of lightning. Small molten globules of metallic iron-nickel and sulfide were expelled from the spinning molten silicate droplets by centrifugal force. The residual silicate droplets crystallized into spheroidal objects called chondrules. Some chondrules collided with others and broke apart. In many cases, whole chondrules and chondrule fragments became entrained in porous dust clumps and were melted again.

Aggregations of chondrules, chondrule fragments, calcium-aluminum inclusions, metal, sulfide, and silicate dust within the disk continued to collide with one another. On average, relative velocities were low enough that gravity eventually caused these aggregations to accrete into porous chondritic planetesimals (figure 1.2). Joining the planetesimals were small amounts of interstellar dust that had formed around dying stars and had been ejected into interstellar space; these materials accreted continuously to the nebula. As the planetesimals grew larger, their gravitational effects increased. The orbits of other planetesimals were perturbed, causing some of these bodies to collide. Such collisions caused the porous planetesimals to compress and collapse.

Most planetesimals were heated, and many were melted. A minority of researchers believe that the porous planetesimals were heated mainly by collisions. Most researchers believe that the

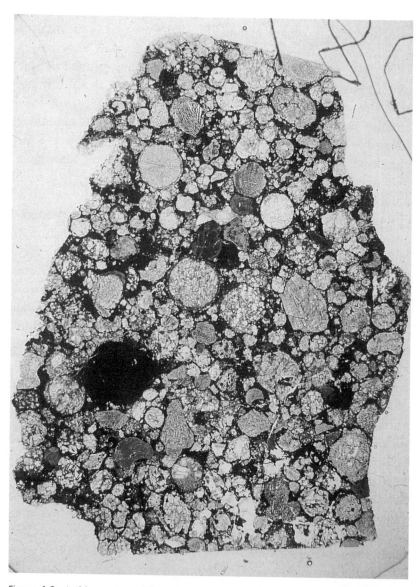

Figure 1.2. A thin section of the unequilibrated ordinary chondrite Semarkona. The rock consists mainly of chondrules and chondrule fragments embedded in a small amount of fine-grained silicate-rich matrix material. The field of view is 13 × 17 mm. (Courtesy of Jeff Taylor, University of Hawaii)

heat came from the decay of the short-lived radionuclide aluminum 26 (i.e., aluminum atoms with 13 protons and 13 neutrons). If sufficient amounts of this isotope were available, the planetesimals would have melted. The aluminum 26 itself may have formed in a recent supernova explosion or through irradiation by the early Sun.

Minor melting along grain boundaries transformed porous materials into rocks. Many of the rocks were heated and recrystallized. In some cases, water-bearing silicate minerals that had accreted to the planetesimals were dehydrated, causing fluid alteration of materials near the surface.

All the planetesimals were hit repeatedly by meteoroids. Some regions were melted by the shock energy released by these impacts; other regions were shattered. In some cases, projectiles colliding at low relative velocities were incorporated intact into the target rocks. Fine-grained material was chipped off rocky outcrops by micrometeoroids to form unconsolidated soil. Rare-gas molecules streaming out of the Sun adhered to the surfaces of tiny mineral grains in the soil. Some grains were damaged by high-energy particles released by solar flares.

In different regions of the nebula, planetesimals colliding at low relative velocities stuck together, some eventually forming the cores of planets. The massive gravity of nascent Jupiter perturbed the orbits of nearby planetesimals, causing them to collide at higher relative velocities. Those bodies with orbital periods resonant with that of Jupiter were flushed out of the region. Many were expelled from the solar system, some fell into the Sun, and others rained down on the rocky planets in the inner solar system. The dispersal of objects in Jupiter's proximity caused a dearth of residual mass in the asteroid belt and in the vicinity of Mars. Mars ended up only 11 percent as massive as Earth.

Because the amount of material in the nebular disk in the vicinity of Uranus and Neptune was probably low, some theorists have argued that these two giant planets did not form in their present locations. Instead, their rocky cores may have formed near Jupiter and Saturn and were then gravitationally perturbed

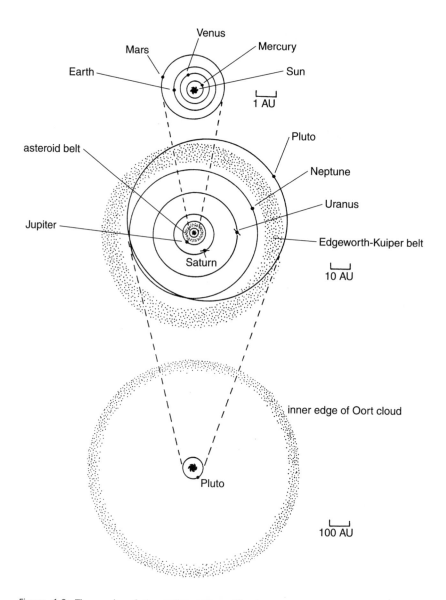

Figure 1.3. The scale of the solar system. The inner planets (rocky and metallic bodies) are in the upper part of the diagram; gas giants and icy planetesimals (including Pluto) of the Edgeworth-Kuiper Belt are in the center; and the inner edge of the Oort Cloud of comets is at the bottom. The scale expands from top to bottom.

away; after a few hundred thousand years, the Uranus and Neptune embryos settled into their present orbits (figure 1.3).

As the giant planets formed, electrically conducting material flowing deep within them interacted with the magnetic field stretching out from the Sun to produce planetary magnetic fields. The conducting material was probably liquid metallic hydrogen in Jupiter and Saturn and an electrolytic solution of water in Uranus and Neptune.

As planetesimals and planetary embryos were scattered about, particularly energetic impacts of these bodies with the major planets tilted the planets and affected their rotation rates. Uranus got tipped on its side by the off-center impact of a projectile the size of Earth. The hypothesized impact on Earth more than four billion years ago of a projectile a few times the mass of Mars probably formed the Moon.

Many planetesimals collided with each other. The remnants now residing between Mars and Jupiter are known as asteroids (meaning "starlike"); these small bodies appear only as pinpoint, starlike images in all but the largest optical telescopes. Some asteroids were hit hard and disrupted, their fragments remaining in orbits similar to those of their parents. These groups of sibling objects possessing similar orbits are known as asteroid families. Other planetesimals were hit hard enough to break apart but not hard enough for most of the fragments to reach escape velocity. The fragments fell back together, forming poorly compacted, chaotic piles of rubble. This process of collisional disruption and gravitational reassembly may have affected most large asteroids, turning them inside out and mixing together materials from the cores and surfaces (figure 1.4).

Some asteroid fragments were left with distinct double-lobed shapes resembling giant peanuts or dog bones. In some cases, these asteroids may consist of two separate bodies in contact. The subsequent collisions of such objects with the Moon or Earth can produce double craters.

As the giant planets formed, small satellites accreted in orbit around them. The planets' more distant satellites were gravita-

impact

disruption

gravitational
reassembly

Figure 1.4. Schematic time sequence showing the process of impact, disruption of the colliding objects, and gravitational reassembly of an appreciable fraction of the debris. Reassembly is defined as occurring when more than half of the fragments of the target asteroid have insufficient kinetic energy to reach escape velocity and, hence, fall back together.

tionally captured as they wandered by. Even Mars (with far less mass than the giant planets) captured a couple of small asteroids in its vicinity. The gravitational fields of the planets caused most of their satellites to become tidally locked, forcing them to rotate on their axes at the same rate that they revolved around their parent planet. Hence, they always show the same face to their primary planet just as the Moon does to Earth.

Gravitational tugs on some of the moons in the outer solar system by their planet and by neighboring moons caused severe tidal heating. Their icy surfaces melted and flowed. Images of Jupiter's moon Europa taken by recent spacecraft flybys have revealed icebergs that appear to have cracked and drifted apart. Most scientists believe that Europa has an ocean of liquid water beneath a cap of ice. Its neighboring moon Io, tugged on by both Jupiter and Europa, experienced great internal heating and expelled much of its initial allotment of volatile compounds from hundreds of volcanoes dotting its surface.

Each of the giant planets produced a system of orbiting rings of dust and ice. Some rings are held in place by the gravitational pulls of small moonlets. Some rings are continuously replenished from particles knocked off nearby satellites by collisions. Some rings may have formed by collisional or tidal disruption of orbiting moons.

Asteroids occupying the same orbital plane as Jupiter but leading or trailing the planet by 60° in their orbits are in gravitationally stable locations called Lagrangian points (the so-called L_4 and L_5 points). Several hundred bodies known as Trojan asteroids are distributed around both of these points. Most were probably captured early in solar-system history (figure 1.5).

Icy planetesimals, commonly known as comets, accreted in the outer solar system where the giant planets now reside. Billions of these planetesimals were flung thousands of astronomical units away from the Sun by gravitational interactions with the giant planets. The Oort Cloud is the spherical swarm of these icy planetesimals at the outskirts of the solar system; it extends out to about 50,000 AU.

Other icy planetesimals remained closer to the Sun, just beyond the orbit of Neptune in a region known as the Edgeworth-Kuiper Belt. The largest of these objects is Pluto. The Edgeworth-Kuiper Belt is also the source of the vast majority of comets entering the inner solar system.

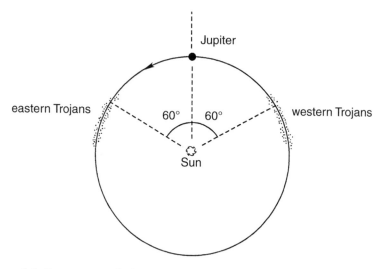

Figure 1.5. Two groups of planetesimals, known as Trojan asteroids, occupy the gravitationally stable Lagrangian points located 60° ahead of and 60° behind Jupiter along Jupiter's orbital path.

A few comets passing through the inner solar system come from the Oort Cloud. As the Sun orbits the galactic center once about every 240 million years, it occasionally passes relatively close to neighboring stars. Gravitational tugs from these stars affect the orbits of comets in the Oort Cloud. The orbits are also affected by the gravitational pull of giant molecular clouds and by the myriad stars in the galactic disk. A few of the Oort Cloud comets are diverted by these gravitational tugs toward the inner solar system. Some occasionally brighten our evening skies, and sometimes one strikes Jupiter, the Moon, or Earth, or explodes in our lower atmosphere.

After Earth had melted, devolatilized, formed a metallic-iron core, and begun to cool, the impacts of icy planetesimals brought water and other volatile compounds to the surface. Sometime around three and a half to four billion years ago, the first organisms emerged from a sterile world. As life evolved on Earth, it had to contend with volcanic eruptions, shifting continents, earthquakes, changing climates, colliding asteroids, and perhaps ionizing radiation from occasional nearby supernova explosions. Early photosynthesizing organisms produced oxygen, and within one and a half to two billion years had oxygenated the ocean and dramatically changed the composition of the atmosphere. As individual organisms lived and died, some minuscule fraction were deposited in sediments that hardened into fossil-bearing rocks. Many of these rocks were later eroded by wind, water, and moving glaciers; others were buried and metamorphosed. But many fossil-bearing rocks remained. Their study has unlocked some of the secrets of the early history of life, although life's ultimate origin has so far proven elusive.

The eventual development of consciousness and speech enabled one species to build telescopes, explore the cosmos, and ponder the possibilities of extraterrestrial life. Out of their brains poured myth, religion, mathematics, technology, philosophy, and science. It was this species, the human species, that first began telling stories of the solar system. But, as the Renaissance philosopher Francis Bacon (1561–1626) cautioned, "[t]he mind of man is like an enchanted glass, full of superstitions, apparitions, and impostures." When filtered through such an imperfect apparatus as the human mind, it is not surprising that our stories tend to reflect our cultural, political, and religious biases.

Nevertheless, our telescopes, spacecraft, and computers have provided enormous amounts of information about the present state of the solar system and its likely origin. Astronomers and solar physicists also have a good idea of the end of the story: in a few billion years, the Sun will balloon out as a red giant, engulfing Mercury and Venus and scorching the surface of Earth. The oceans will boil, and life on Earth will cease. While many details of solar-system history remain hidden, we have enough information to spin a good yarn.

A fundamental property of the solar system is its position in the Galaxy. Surprisingly perhaps, this location is important to the history of life. The center of the Galaxy is an active region, full of radiation. If the solar system had formed near there, the origin of life would have been problematic. On the other hand, stars in clusters above the plane of the Galaxy are poor in heavy elements and probably lack rocky planets. The story of how scientists deduced the location of the solar system is an instructive one in the history of astronomy. It is filled with brilliant insights, painstakingly obtained data, numerous mistakes, and agonizingly slow progress. It is the story we take up next.

II

WHERE ARE WE?
THE LOCATION OF THE
SOLAR SYSTEM

O

Death in itself is nothing; but we fear
To be we know not what, we know not where.
—John Dryden, *Aureng-Zebe*

THERE IS A country on Earth called Comoros. Most people have not heard of it, but if asked to write down things they would like to learn about the country, they might list size, climate, topography, population, and language. They may be interested in its economy, history, dominant religion, and form of government. But there is one thing that would appear on everyone's list—location. Just where in the world is Comoros? Is it a landlocked country nestled between India and Tibet? (That would be Bhutan.) Is it east of Guatemala on the shores of the Caribbean? (Belize.) Or is it a group of islands in the Mozambique Channel off the southeastern coast of Africa? (Bingo.)

Just as it is with a country on Earth, so it is (or should be) with the solar system. Just where in the Galaxy is it? And how did scientists figure this out?

From our vantage point at Earth's surface, we speak of sunrise and sunset, echoing the ancient belief that the Sun revolves around the Earth. Such a geocentric cosmology was embraced by Aristotle in the fourth century B.C.E. and was assumed by the writer of Joshua 10:13, "[s]o the Sun stood still in the midst of heaven and hasted not to go down about a whole day."

The first heliocentric model was formulated in the third century B.C.E. by Aristarchus of Samos, a Greek philosopher who worked in the great library in Alexandria, Egypt. Aristarchus observed that Earth cast a large shadow across the Moon during lunar eclipses and reasoned that the Sun must therefore be much larger than Earth and quite distant. Probably because it seemed philosophically inelegant for a large body to go around a smaller one, Aristarchus scrapped the geocentric model and proposed that Earth rotated once each day (to account for the daily rising and setting of the Sun) and revolved around the Sun once each year (to account for the seasonal changes in the constellations). Aristarchus's model marked the first inkling of the existence of the solar system—the system could not have been considered *solar* when the Sun was believed to be just a small luminous body revolving around Earth.

However, the heliocentric model foundered on the basis of its denial of common sense (it sure looks as if the Sun goes around Earth), its contradiction of Aristotelian physics (didn't the old sage know everything?), and its failure to conform to observational evidence. The evidence against heliocentrism was the failure to observe parallax—the apparent change in position of the stars as seen from Earth over the course of half a year. (Parallax is an everyday phenomenon, known to the ancient Greeks. If you hold your thumb up at arm's length and close one eye at a time, you will see the apparent shift in the position of your thumb relative to background objects.) If the stars were relatively close by, as the Greeks believed, then a parallactic shift should be observed if Earth went around the Sun. The shapes of the constellations would even appear distorted between one season and the next. The absence of this shift constituted powerful evidence against the heliocentric model. (Of course, the reason that parallax was not observed is because the stars are too far away for this effect to be seen with the unaided eye — the nearest star is about 265,000 times farther away than Earth is from the Sun.)

In the ensuing years, Aristarchus's model was generally ignored, and most of his writings were lost. Geocentrism ruled the day. Its greatest exposition was made by Claudius Ptolemaeus (better known as Ptolemy) about 140 C.E. He also worked in Alexandria and had access to the great library. He developed a sophisticated theoretical model that predicted the positions of the planets to within 5°. In his book, known today by its Arabic transliteration as the *Almagest*, Ptolemy paid homage to Aristotle and assumed, like his predecessors, that the planets, Sun, and Moon moved around Earth in circular orbits. To account for the apparent nonuniform motion of the planets, Ptolemy modeled these bodies as each moving around a small circle (an epicycle) whose center was on the circumference of a larger circle (a deferent) that moved around Earth. Earth itself was displaced from the center of the concentric deferents that carried the planets (figure 2.1).

Ptolemy's model remained the accepted description of planetary motion for more than fourteen hundred years. Over the centuries, the Ptolemaic system was refined by Arabic and European astronomers. In the thirteenth century C.E., King Alfonso X of Castile sponsored a revision of the system and published a set of tables of calculated planetary positions that better matched their observed positions. The *Alfonsine Tables* were widely circulated. In the sixteenth century one copy reached a young church canon in Poland named Nicolaus Copernicus. Copernicus had studied astronomy, law, and medicine and was eager to make astronomical observations of his own. He noted that a conjunction (an apparently close passage in the sky) of Jupiter and Saturn in 1504 occurred ten days off from the date predicted by the *Alfonsine Tables*. This inaccuracy troubled Copernicus, as did the inelegant

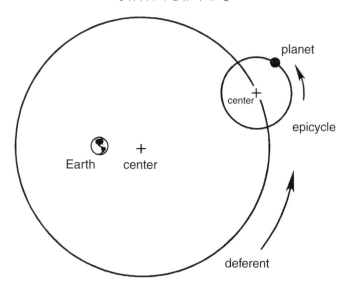

Figure 2.1. In the Ptolemaic view of the solar system, planets move around a small circle called an epicycle. The center of the epicycle is on the circumference of a large circle called a deferent that itself moves around Earth (which is offset from the center of the deferent).

displacement of Earth from the center of the deferent in the Ptolemaic model.

Copernicus sought to reintroduce uniform circular motion to the planets and developed a new heliocentric model. He explained away the failure to observe stellar parallax by assuming that the stars resided in a spherical shell that was at a distance greater than the Earth–Sun distance. Although he had tried to make his model simpler than Ptolemy's, Copernicus was forced to add small epicycles to the orbits of the Sun, Moon, and planets to account for minor variations in their observed movements. The complete model of Copernicus was published as *De Revolutionibus Orbium Coelestium* in 1543, the year Copernicus died.

Erasmus Reinhold, an astronomer at the University of Wittenberg, used the Copernican system to calculate planetary positions. These were published by almanac makers and came to be known as the *Prutenic Tables*. Although it was no better at predicting planetary positions than Ptolemy's model, the Copernican hypothesis and the *Prutenic Tables* planted the seeds for a revolution in astronomical thought.

Within seventy years of Copernicus's death, Galileo used the newly invented telescope to discover four moons circling Jupiter. It became clear that the Aristotelian idea that Earth was the center of all heavenly motion was wrong. Furthermore, in those days before Newton's explanation of universal gravitation, critics of the heliocentric model had asserted that if Earth really moved around the Sun, it would leave the Moon behind. Galileo found that as Jupiter moved through the heavens, it retained its retinue of moons; he reasoned that if Jupiter could do it, so could Earth.

Four additional astronomical discoveries of Galileo chipped away at the foundations of the Ptolemaic system. First, Galileo reported that the Moon had valleys and mountains and hence was not a perfect, unblemished body. This observation removed a philosophical impediment. Imperfect bodies need not be expected to move with perfectly uniform circular motions. Second, he observed sunspots, thereby casting doubts on the Sun's perfection. By recording the movement of sunspots across the face of the Sun, Galileo recognized that the Sun itself was a sphere rotating on its axis. If the Sun could rotate on its axis, so could Earth. Third, he observed that Venus showed phases just like the Moon. In the Ptolemaic model, Venus was located on an epicycle between Earth and the Sun and would always appear as a crescent (figure 2.2). In the Copernican model, Venus orbited the Sun and should appear as a crescent only when its orbit took it between Earth and the Sun. When it was on the other side of its orbit, Venus should appear full or gibbous, just as Galileo observed. By 1613 Galileo was able to write with confidence that "Venus revolves around the Sun just as do all the other planets." Finally, Galileo found that the Milky Way was composed of innumerable stars too faint to see with the naked eye. If stars were all about the same size, then the faint stars in the Milky Way were very far away. No longer could the stars be considered confined within a thin spherical shell revolving around Earth, as Ptolemy had assumed.

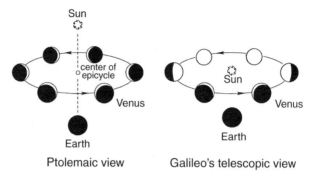

Figure 2.2. In the Ptolemaic model (*left*), Venus would always appear from Earth as a crescent. In the Copernican model (*right*), Venus would show phases like the Moon. Galileo's small telescope revealed that Venus had phases.

Although Galileo's discoveries provided strong evidence for the heliocentric model, the model was still unable to provide accurate predictions of planetary positions. This was because Galileo had continued to assume that planetary orbits were circular. However, the path toward the solution of Galileo's dilemma had actually been blazed decades earlier by a Danish nobleman named Tycho Brahe, who had dedicated his life to obtaining accurate planetary positions.

In August 1563, the year before Galileo was born, Tycho observed a close conjunction of Jupiter and Saturn and noted that its prediction was off by a month in the *Alfonsine Tables* and by several days in the *Prutenic Tables*. Determined to make more accurate planetary observations in those pretelescopic days, Tycho constructed large naked-eye observing instruments and recorded the positions of the Sun, Moon, and planets nearly every day for two decades. In 1600 he hired a young German astronomer, Johannes Kepler, to help him prepare more accurate tables of planetary positions. These were to be called the *Rudolphine Tables* after Tycho's patron, Holy Roman Emperor Rudolph II.

However, Tycho died in 1601, and Kepler acquired the books containing Tycho's meticulous planetary observations. Over the next eighteen years, Kepler used these data to derive his three laws of planetary motion: (1) *The orbits of the planets around the Sun are ellipses with the Sun at one focus.* This law destroyed the Greek concept of circular planetary motions. (2) *A line from a planet to the Sun sweeps over equal areas in equal times,* that is, planets move faster when closer to the Sun in their orbits than when farther away. This law destroyed the concept of uniform planetary motions. (3) *The square of a planet's orbital period is proportional to the cube of its heliocentric distance.* This law enabled astronomers to calculate a planet's distance from the Sun (relative to Earth's distance) from the planet's observed orbital period.

In 1628 Kepler published the *Rudolphine Tables* of planetary positions at his own expense and dedicated them to Tycho's memory. These tables were the first to predict planetary positions accurately.

Isaac Newton's publication of *Principia Mathematica* in 1687 demonstrated that a single force — gravity — is responsible for the regular motions of solar-system bodies. This insight was confirmed in spectacular fashion by the predicted reappearance of a great comet in 1758 after Edmond Halley had intuited that the bright comets of 1531, 1607, and 1682 were caused by a single body in an approximately 76-year orbit around the Sun. These studies served to establish the fundamental interconnectedness of the solar system.

Although Copernicus, Galileo, and Kepler had demonstrated that Earth was not the center of the universe, seventeenth-century astronomers had no doubt that the Sun was at the center. This view persisted for nearly three hundred years.

In the late eighteenth century, Sir William Herschel and his sister Caroline Herschel attempted to determine the three-dimensional extent of the Milky Way by counting visible stars in different directions. Using a 48-inch telescope (at the time the largest in the world) that Sir William had constructed, the Herschels counted stars in 683 different directions. They assumed that the Sun was located within a great cloud of stars and that fainter stars were farther away than brighter stars. If they found few faint stars in a particular direction, it meant that the edge of the cloud was nearby; if faint stars were numerous in a particular direction, it meant the edge in that direction was distant.

The Herschels concluded from the many faint stars in the Milky Way that the star system was an irregularly shaped disk extending far in the Milky Way's direction but not very far in the direction perpendicular to it. They called their inferred amoeba-shaped structure the "grindstone model" of the Milky Way, after the stone used in mills to grind flour. The solar system appeared to be near the center of the disk, but the absolute dimensions of the disk were unknown. William Herschel estimated that the diameter of the Milky Way was at least nine hundred times the distance to Sirius, the brightest star in the sky.

Herschel also observed many spiral-shaped nebulous objects in the night sky. Although his telescope could not resolve individual stars within them, he concluded that the spiral nebulae were other disk-shaped star clouds like the Milky Way. He surmised that from a great distance, the Milky Way would look just like them. Herschel had been anticipated by several decades in this view by the English theologian Thomas Wright and the German philosopher Immanuel Kant. Kant had proposed that the spiral nebulae be called "island universes," a picturesque term that put these objects on par with the Milky Way.

Other nineteenth-century astronomers duplicated the Herschels' work and came to similar conclusions about the Milky Way being approximately centered around the Sun. This effort culminated in a series of studies by the Dutch astronomer Jacobus C. Kapteyn in the early twentieth century. He determined the number of stars in different directions as well as their brightness and motion. He used statistical techniques to find the average distance of the stars in the Milky Way and was the first to derive a numerical estimate for the size of the Galaxy. Kapteyn concluded that the Sun was within 8,000 light-years of the center of a disk 26,000 light-years in diameter and 6,500 light-years thick. Kapteyn was concerned, however, that if starlight is absorbed in space, then faint stars may be closer to the Sun than they appear and distant stars might not be visible at all. Appreciable absorption of starlight would mean that the Galaxy was larger than Kapteyn's estimate, but it was not until about 1930 that Robert J. Trumpler at Lick Observatory near San Jose, California, confirmed the absorption of starlight by interstellar dust.

The next chapter in the discovery of the position of the solar system in the Milky Way began with the work of American astronomer Henrietta S. Leavitt in 1912. She was studying variable stars in the Magellanic Clouds (two small irregular satellite galaxies of the Milky Way) and found that for a particular type of variable (now known as Cepheids after the star Delta Cephei in our own Galaxy), there was an important relationship between luminosity (or brightness) and the period of time between episodes of maximum brightness: the brighter the star, the longer the period. Leavitt's work was refined by Danish astronomer Ejnar Hertzsprung and Harlow Shapley, an astronomer at Mt. Wilson Observatory (located in the San Gabriel Mountains near Los Angeles). These astronomers calibrated the period-luminosity relationship. Shapley found that another class of variable star, the RR Lyrae stars, had a different period-luminosity relationship.

In 1918 Shapley published a paper on his studies of the distributions of star clusters in the Galaxy. He noted that the so-called "open clusters" such as the Pleiades and Hyades (which typically consist of a few dozen to a few hundred loosely packed stars) are confined to the disk of the Milky Way. However, the "globular clusters" (which typically consist of densely packed spheroidal associations of tens of thousands to hundreds of thousands of stars) do not occur in the disk at all. More than half lie within or close to the constellation Sagittarius. Shapley hypothesized that the globular clusters are gravitationally bound to the Galaxy and orbit about the center. In that case they should be uniformly distributed around the center. Their observed locations indicated that the center of the Galaxy was in the direction of Sagittarius; the Sun was consequently shunted off to the galactic outskirts (figure 2.3).

Shapley found variable stars within the globular clusters and, from the period-luminosity relationship, determined both their

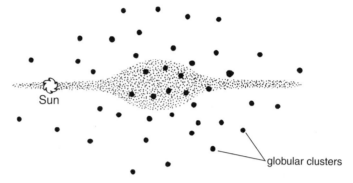

Figure 2.3. Cross-sectional view of the Galaxy showing the distribution of globular clusters. From the Sun's vantage point near the galactic outskirts, most globular clusters appear in one region of the sky, toward the galactic center in Sagittarius.

distance from Earth and the overall dimensions of the Galaxy. Shapley's Galaxy was ten times larger than Kapteyn's, a full 300,000 light-years in diameter. This value is more than three times larger than modern estimates. The error occurred because Shapley had mistaken the RR Lyrae variable stars in the globular clusters for Cepheids and had consequently used the wrong period-luminosity relationship.

At the beginning of the twentieth century, the nature and distances of the spiral nebulae were still in dispute. Some researchers like Shapley thought them to be clouds of gas at the edges of the Milky Way. Some thought, as Herschel, Wright, and Kant had more than a century earlier, that they were distant galaxies resembling the Milky Way. An important step toward resolving this issue was made by astronomer George W. Ritchey of Mt. Wilson Observatory, who discovered a nova in one of the spiral nebulae in July 1917. Within two months, astronomers searching old photographic plates discovered ten more novas in spiral nebulae. Because a nova is a dying star, these discoveries strongly supported the idea that the spiral nebulae were star systems like the Galaxy rather than clouds of interstellar gas.

The controversy came to a head in April 1920, when Shapley and Heber D. Curtis of Lick Observatory debated the opposing positions at a meeting of the National Academy of Sciences in Washington, D.C. Curtis argued forcefully that the spiral nebulae were distant galaxies. He suggested that the large variation in apparent diameter of the spiral nebulae indicated that they resided at different distances and thus could not all be within the Milky Way. The apparent absence of spiral nebulae in the galactic disk was taken to mean that these are distant objects whose light is obscured by interstellar material. He also pointed out that the spectral signatures of the spiral nebulae were like agglomerations of stars (dark lines on a bright background) and unlike those of thin clouds of gas (bright lines on a darker background). The point was conceded by the Shapley camp after 1924, when photographs made by Edwin Hubble with the new 100-inch (2.5-meter) telescope on Mt. Wilson resolved individual stars in the great spiral nebula in Andromeda. Hubble found that some of the stars were Cepheid variables; their period-luminosity relationship indicated that the Andromeda nebula was far beyond the Milky Way.

Within five years of the circa 1930 discovery by Robert Trumpler of the absorption of starlight by interstellar dust, astronomers were making fairly accurate estimates of galactic and intergalactic distances. As the years went by, these estimates were refined. The modern value for the distance to the Andromeda Galaxy is 2,180,000 light-years.

The recognition that the spiral nebulae were galaxies in their own right, much like the Milky Way, enabled astronomers to perceive galactic shapes and study the complexity of spiral structure. The spiral arms of galaxies are home to hot, bright O and B stars, young open clusters, clouds of neutral hydrogen (known to astronomers as H I regions), and clouds of ionized hydrogen (known as H II regions). From such observations, astronomers inferred the spiral pattern of our own Galaxy, and by the 1950s began mapping the galactic spiral structure in the solar neighborhood (figures 2.4 and 2.5).

Because interstellar dust is concentrated in the spiral arms, our view of the galactic disk is obscured by the absorption of starlight

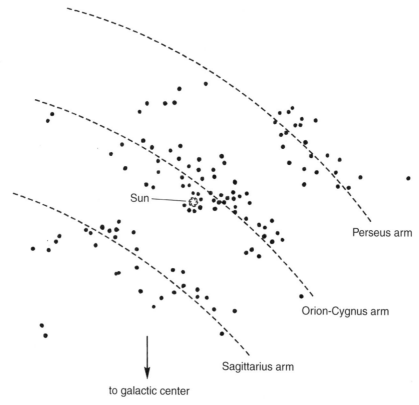

Figure 2.4. The very bright and very young O and B stars tend to occur together in stellar associations that mark the locations of the spiral arms of the Galaxy. The O and B associations in the solar neighborhood trace out three spiral arms: the Orion-Cygnus arm (in which the Sun is located), the Sagittarius arm (located toward the galactic center), and the Perseus arm (located in the opposite direction). (Diagram after fig. 16-16 of M. A. Seeds (1999), *Foundations of Astronomy*)

at visual wavelengths. However, radio wavelengths are much longer than the size of the dust particles and are thus not absorbed. Observations of radio signals with a wavelength of 21 centimeters (produced by neutral hydrogen) allow radio astronomers to map the distribution of H I clouds in the galactic disk. The radio emission of carbon monoxide (CO) can be used to locate giant molecular clouds in the disk. Astronomers have com-

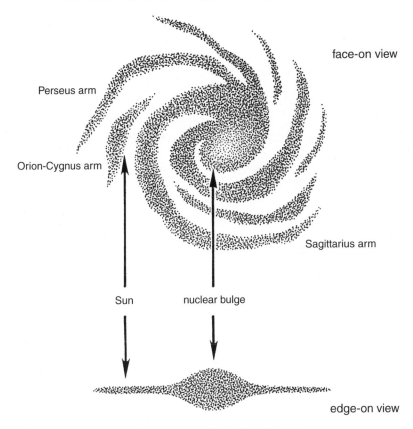

face-on view

Perseus arm

Orion-Cygnus arm

Sagittarius arm

Sun   nuclear bulge

edge-on view

Figure 2.5. Face-on and edge-on views of the Milky Way Galaxy.

bined radio and optical observations to map the local spiral structure. They have found that there are three distinct spiral arms in the Sun's vicinity. Each is named after the constellation in which prominent features of the arm are located. The solar system is situated along the inner edge of the Orion-Cygnus arm within a small spur bent inward toward the galactic center. Beyond us, away from the center of the Galaxy, is the Perseus arm; closer in is the Sagittarius arm (figure 2.6). Some astronomers have inferred the existence of a fourth arm (the Centaurus arm) even closer to the galactic center.

Like all spiral galaxies, the Milky Way (figure 2.7) rotates

Figure 2.6. Photograph of the center of the Galaxy in Sagittarius made by wide-angle 35-mm camera. Also visible is the trail of the *Echo* satellite. (Courtesy of U.S. Naval Observatory)

Figure 2.7. The Milky Way Galaxy in infrared light as revealed by the Diffuse Infrared Background Experiment of the COBE (Cosmic Background Explorer) satellite. (Courtesy of Ned Wright, UCLA)

around the galactic nucleus with its arms trailing. The Sun takes about 240 million years (one galactic year) to go around the Galaxy once.

The final piece of information required to pinpoint the location of the solar system is an accurate value for the Sun's distance from the center of the Galaxy. Methods of determining this distance include the direct measurement of the motion across the sky of objects near the galactic center, finding the distances to objects assumed to be symmetrically distributed around the galactic center (as Harlow Shapley had done with globular clusters), and determining the galactic rotation period of variable stars and molecular clouds of known distance. In 1993 Mark J. Reid of the Harvard-Smithsonian Center for Astrophysics summarized the methods and obtained a weighted average value of 26,000 ± 2,000 light-years for the Sun's distance from the galactic center.

In the twenty-three centuries since Aristarchus of Samos first introduced the heliocentric model of the solar system, we have come to know where in the Galaxy the Sun resides. As we gained this knowledge, our view of the universe shifted from that of a small system centered on Earth to that of an immense expanding structure containing more than a hundred billion galaxies. Now, after thousands of years of astronomical observations, we have come to know our place in the cosmos.

Part of our new understanding of the universe involves a more detailed knowledge of the features of the solar system. More than a list of planets, moons, comets, and asteroids is required. It is essential that we explore the energetic interactions among solar-system bodies. Because many of these interactions involve the transfer of heat from one body to another or the transformation of kinetic or chemical energy into heat, it is important to explore the sources of heat in the solar system. Although the Sun is the most obvious heat source, it is not the only one. Tidal effects induced by gravitational tugs and collisions caused by errant asteroids and comets are additional sources. This topic is examined in the next chapter.

III

HEAT SOURCES

O

Fear no more the heat o' the sun,
Nor the furious winter's rages;
Thou thy worldly task hast done,
Home art gone, and ta'en thy wages;
Golden lads and girls all must,
As chimney-sweepers, come to dust.
　　　　—William Shakespeare, *Cymbeline*

IF YOU were to stand on Pluto, the solar system would seem like a cold place. The bright star in the sky provides little heat, no volcanoes protrude above the planet's icy surface, and in that frigid world there is not a single fire. The Plutonian perspective is, however, a limited one; heat sources abound in the solar system. Some are active today; others were active four and a half billion years ago, when the solar system began to form. Without reliable sources of heat, there would be no life.

We will examine the solar system's energy sources by taking a trip to the southwestern United States. We'll start our tour on the eastern rim of Meteor Crater in northern Arizona. Grab a good pair of binoculars; bring along a lawn chair and some sunblock.

Figure 3.1. A 1900 ¼-real coin from Guatemala showing the major heat source of the solar system—the Sun—and an erupting volcano. The volcano is a manifestation of heat produced by the decay of radioactive elements in Earth's interior.

It is late morning on a sunny day in mid-August. Before us is the crater, deeper than the Washington Monument is tall and more than 1¼ kilometers across. Behind us, to the east, is the quarter Moon rising over the Painted Desert. To the northwest is the San Francisco Volcanic Field. Above us is the Sun (figure 3.1).

As the Sun climbs higher in the sky, we can feel its rays warm our skin. Glancing up quickly, we glimpse a brilliant yellow-white ball. The Sun is the most prodigious source of heat in the solar system. Fusion, the transmutation of hydrogen nuclei into helium nuclei, takes place in the dense central core of the Sun. Energy is released from the core in the form of x-rays and gamma rays, but in the surrounding region, this radiation can travel only about 1 centimeter before being absorbed and reemitted. Over the course of tens of thousands of years, the radiation gradually loses energy as it makes its way toward the Sun's surface. Energy is eventually deposited below the surface in hot bubbles of gas that rise convectively to the surface and radiate their heat. Eight minutes later the Sun's rays reach us.

The mean temperature at Earth's surface is 22°C, but if it were not for the small amount of carbon dioxide in the atmosphere (about 0.032 percent), the mean temperature would be much colder, only −13°C. This difference arises from the greenhouse effect. Earth's atmosphere is transparent to the visible radiation from the Sun; the Sun's rays pass through the atmosphere and heat the surface. The warmed surface then radiates its own heat at infrared wavelengths. Instead of escaping into space, however, this infrared radiation is absorbed by carbon dioxide and water vapor; atmospheric heating results. The atmosphere then radiates heat back to the surface.

A far more pronounced greenhouse effect occurs on Venus, where carbon dioxide, water vapor, and sulfuric acid droplets in the dense atmosphere cause the surface temperature to rise by about 375°C. Thus, although the Sun is the ultimate heat source, the greenhouse effect is responsible for significantly raising the surface temperatures of the two largest terrestrial planets.

The Sun is not the only heat source in the solar system. The volcanoes, cinder cones, and lava flows to the northwest of Meteor Crater demonstrate that there is another source of heat deep within Earth (figure 3.2). This heat is due mainly to the decay of four radioactive isotopes within the rocks of the crust and mantle: uranium 235, uranium 238, thorium 232, and potassium 40. Not all potassium is radioactive; only 0.01 percent of naturally occurring potassium undergoes radioactive disintegration.

The half-lives of these isotopes are very long, ranging from 700 million years for uranium 235 to 14 billion years for thorium 232. These isotopes disintegrate spontaneously, emitting helium nuclei, gamma rays, and electrons. The concept of a "half-life" refers to the fact that half of the atoms in a given amount of a radioactive isotope transform into more stable atoms in a set period of time. For example, after 700 million years, a sample con-

Figure 3.2. Castle Geyser in Yellowstone National Park ejecting a column of hot water and steam. Geysers form when groundwater is heated and some turns into steam upon encountering hot rocks or magmatic bodies beneath the surface.

taining 1000 atoms of uranium 235 would have transformed into a sample containing 500 atoms of uranium 235 and 500 atoms of lead 207; after another 700 million years, the sample would contain 250 atoms of uranium 235 and 750 atoms of lead 207. And at the end of three half-lives the sample would contain 125 atoms of uranium 235 and 875 atoms of lead 207.

The energetic particles emitted during radioactive decay are absorbed by the surrounding rock, and the kinetic energy of these particles is transformed into heat. As the rocks heat up, their constituent atoms vibrate more rapidly. In so doing, they jostle adjacent atoms. This process continues as the heat energy makes its way toward the surface in a process known as conduction.

Convection, on the other hand, is a more efficient process of heat transfer. Hot, deeply buried material flows very slowly toward the surface, releases its heat, and very slowly sinks again. At those locations in the upper mantle where the temperature exceeds that necessary for melting, magma chambers form.

When we turn our binoculars toward the Moon, we can make out the maria — the dark lava-filled impact basins that dominate the near side and form the facial features of the Man in the Moon. Long-lived radionuclides are probably responsible for melting the rocks of the lunar mantle and creating these lava flows (figure 3.3). Because of its small size, however, the Moon experiences a high rate of heat loss from its surface. The abundances of uranium, thorium, and potassium are too small to have allowed the Moon to have remained volcanically active up to the present. The youngest lunar lavas are probably more than two billion years old.

Remote sensing of Vesta, the second largest asteroid, show that its surface is also covered with dark lava. Vesta is 250 times smaller than the Moon, and heat is lost from its surface so quickly that long-lived radionuclides are absolutely ruled out as an adequate source of heat. Most meteorites also show evidence of having been heated; some have been melted. Because the vast majority of meteorites are pieces of asteroids, most must come from bodies even smaller than Vesta. Another source of heat

Figure 3.3. Vesicular basalt 15016 from the Moon picked up by the *Apollo 15* astronauts at Hadley Rille. The vesicles (cavities) formed from gas bubbles in the original lava flow from which the rock crystallized. The cube at lower right is 2 cm on a side. (Courtesy of NASA)

must be responsible for heating asteroid-size bodies. One primary candidate is the decay of the short-lived radionuclide aluminum 26. Another is by impacts of projectiles appreciably larger than the one that formed Meteor Crater (figure 3.4). A third candidate is heating by electrical currents induced in asteroid-size bodies by a fierce solar wind of partially ionized gas. Such powerful solar winds were probably emitted by the Sun when it was a young star. These potential heat sources are explored in a later chapter.

As the afternoon wears on and we get hotter, we imagine that we are walking along the beach in southern California, 700 kilometers to the west. At low tide, the beach is wide and we can sometimes find seashells. At high tide, the beach narrows, and no

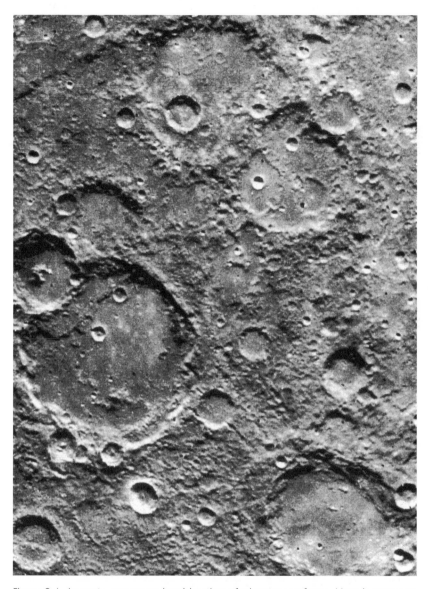

Figure 3.4. Impacts can cause local heating of planetary surfaces. Here impact craters are shown from the south polar region of Mercury, photographed by the *Mariner 10* spacecraft from a distance of 57,300 km. The area shown is 370 × 480 km. (Courtesy of NASA)

shells can be found. As the Moon climbs higher in the sky, we recall that ocean tides are caused by the gravitational tugs of the Moon and the Sun. Just as the Moon raises tides in Earth's oceans and solid matter, Earth raises tides in the Moon. When the Moon is closest to Earth, tidal bulges within the Moon trigger moonquakes at depths of 800 to 1,100 kilometers. Slight frictional heating of the lunar interior results.

Much more significant tidal heating occurs within the Galilean satellites of Jupiter. The two outermost of these moons, Ganymede and Callisto, have ancient ice-rich surfaces covered with numerous impact craters. Europa, the next moon inward, has icy plains and very few impact craters; fractures crisscross its surface. The scarcity of craters indicates that Europa's surface is young. Its icy crust must have been melted and refrozen after the intensive bombardment of meteoroids was over. Tidal stresses induced by nearby massive Jupiter supply the heat that melted the ice. Io, the innermost Galilean moon, is the most tidally stressed body in the solar system (figure 3.5). With Jupiter tugging on one side and Europa on the other, Io goes from high tide to low tide every 42 hours; its surface rises and falls by up to 90 meters. Such flexure generates intense heat in Io, causing widespread internal melting. Ascent of the molten material to the surface leads to the construction of volcanic landforms, hot lava lakes, towering 16-kilometer-high mountains, and exploding volcanoes. During spacecraft flybys, volcanoes on Io were photographed ejecting plumes of debris nearly 300 kilometers above the surface.

Late in the afternoon, we notice storm clouds rolling in from the west. As the clouds block the Sun, we appreciate the cool breeze, but soon it starts to rain. As the storm builds, we see lightning strike the ground in the distance. If we were later to investigate this patch of ground, we might find that the heat from the lightning bolt had melted the quartz sand, forming long hollow tubes of siliceous glass called fulgurites (figure 3.6). There is a massive, discontinuous fulgurite in Michigan that extends for 30 meters. Temperatures exceeding 1700°C were reached in the soil during the lightning strike that formed this object.

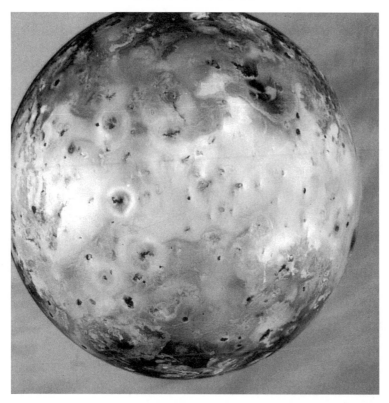

Figure 3.5. Io is the most tidally stressed and most volcanically active body in the solar system. Io as seen here is in front of Jupiter's clouds, photographed by the *Galileo* spacecraft from a distance of 487,000 km. (Courtesy of NASA)

Lightning is an extremely common phenomenon: there are approximately one hundred lightning blasts each second on Earth. However, lightning is not restricted to our planet; it also occurs in the turbulent atmospheres of Jupiter and, perhaps, Venus. A number of astrophysicists have theorized that lightning was also active in the early solar system. More than four and a half billion years ago, the solar system consisted of an infant Sun surrounded by dust (mostly concentrated in a central plane) and overlying gas. The dust revolved around the Sun at velocities appropriate for solid particles at their particular heliocentric distance. The gas revolved more slowly because gas pressure tended to push the gas

Figure 3.6. A hand-held fulgurite formed from quartz sand by a lightning strike. (Courtesy of Paul Roques, Griffith Observatory)

radially outward, away from the Sun. This difference in rotational velocity between the gas and dust may have produced enough turbulence at the dust-gas interface to have caused electrons and ions to separate. Discharges by lightning bolts would have resulted. Porous dust grains that were intercepted by these bolts may have melted. This is one explanation for the formation of chondrules—submillimeter-size solidified droplets of silicate melt that constitute up to 70 percent of the most common meteorites. Other plausible chondrule-formation mechanisms include heating by solar-flare protons and by high-energy electrons accelerated by an intense interplanetary magnetic field.

As evening comes, the storm clears, and we await the spectacle of the annual Perseid meteor shower. If the Moon were not so bright, we would expect to see more than sixty meteors per hour. A meteor is the luminous phenomenon resulting from the passage of a small particle (a meteoroid) through Earth's atmosphere. Most such particles appear to be cometary debris. As each particle plunges at cosmic velocity through the tenuous outer atmosphere, friction heats the object and causes an electric charge to

Figure 3.7. Jupiter and its Great Red Spot as photographed by the *Pioneer 11* spacecraft from a distance of 1,063,000 km. (Courtesy of NASA)

build up in the surrounding air. This results in a bright glow. An analogous mechanism must have operated in the early solar system. As gas and dust revolved around the infant Sun, interstellar grains, attracted by the Sun's gravity, fell into the solar system at cosmic velocities. The grains were rapidly decelerated by gas drag. If there were sufficient numbers of grains, the resulting frictional heat could not have been radiated away fast enough, and the grains would have melted. Such a mechanism has been occasionally invoked by researchers to account for the formation of chondrules and other inclusions in meteorites.

As we watch the night sky, our gaze falls upon Jupiter (figure 3.7), which appears in our binoculars as a bright and slightly oval disk. Jupiter began its existence as a diffuse cloud of gas millions of kilometers across. Over tens of thousands of years it gradually collapsed under its own gravity to form a much smaller proto-planet, still several times Jupiter's present size. Jupiter continues to contract today, but at a much slower rate. The energy released by the conversion of gravitational potential energy to heat generates powerful convective currents inside the planet. Upwelling currents of warm gas produce the bright zones that wrap around the planet's disk. Sinking regions of cooler gas form the parallel dark belts. So much internal heat is generated that Jupiter currently emits twice as much infrared radiation as it receives from the Sun. Saturn experienced a similar but less extreme history. It currently radiates about 45 percent more infrared energy than it receives from the Sun.

High clouds drift in and obscure our view of the sky; it is time to leave. As we drive along I-40, we remember the clogged drain in the bathtub at home and decide to stop at the market for some Drāno. The addition of a sodium hydroxide product to the standing water in the tub causes an exothermic chemical reaction as the sodium hydroxide goes into the solution. Astrophysicist Donald Clayton once suggested that exothermic chemical reactions within aggregates of interstellar molecules were responsible for chondrule formation during the collapse of a dense molecular cloud to form the solar system. Warming of the aggregates by heat generated by the collapsing cloud may have led to exothermic reactions by unstable molecules. These reactions in turn may have triggered additional exothermic reactions, eventually culminating in the melting of the aggregates. Although most researchers do not believe that chondrules formed this way, it is not out of the question that a fraction of the dust in molecular clouds are heated by this mechanism.

This diverse array of heat sources is responsible for the complexities of the objects that populate the solar system. One of the main tasks of geoscientists and space physicists is to infer the thermal histories of an astounding variety of objects from chondrules to asteroidal basalts, from the Sun's core to the satellites of the outer planets, from the beginning of solar-system history to the distant future. It is not an easy task, but it is an exciting one.

Part 2

EARTH AND MOON

ASTRONOMERS and planetary scientists sometimes lose sight of the fact that Earth is the most interesting body in the solar system. Venus, Mars, and Io may have volcanoes, Mars may once have had a small ocean, the giant planets may have intense magnetospheres, but no planetary body has such a diverse array of structures as Earth. A glance at the contents of undergraduate geology and ecology textbooks reveals such terrestrial phenomena as oceans, glaciers, deserts, rivers, volcanoes, converging and diverging lithospheric plates, grasslands, rain forests, limestone caves, alpine forests, impact craters, atolls, and sand dunes. We can read about monsoons and tornadoes, landslides, earthquakes, shoreline erosion, weathering, lightning, and geomagnetism. It is hard to avoid being impressed by the incredible variety of igneous, metamorphic, and sedimentary rocks. Most interesting of all is the curious phenomenon of life. We know of no other place where living organisms can be found.

The history of Earth is interwoven with the history of the Moon. Most planetary scientists believe that the Moon formed from Earth more than four billion years ago during a giant impact. Since its formation, the Moon has tugged at Earth, slowing its spin; as a consequence of the law of conservation of angular momentum, the Moon's orbit has widened.

Earth is also an integral part of the solar system. Gravitational tugs of the Sun and Moon on Earth periodically change the inclination of its spin axis and the shape of its orbit. These changes affect the amount of sunlight reaching Earth's surface and appear to be responsible for the onset of periods of extensive glaciation.

The discovery of the nature of Earth's magnetic field is another instructive lesson in the history of science. It spans more than two thousand years from the invention of the compass to the discovery that Earth's field occasionally switches polarity. A deep understanding of Earth's magnetic field would have many applications beyond Earth: solar magnetic phenomena (including the sunspot cycle and solar flares), the magnetic fields of other planets, the extremely intense magnetic fields of rapidly rotating neutron stars, and the magnetic alignment of elongated dust grains in the interstellar medium. Spontaneous reversals of Earth's magnetic field are local manifestations of a truly universal phenomenon. By studying magnetic Earth, we advance our understanding of the magnetic cosmos.

IV

THE MAGNETIC EARTH

O

I am he that walks with the tender and growing night,
I call to the earth and sea half-held by the night.
Press close bare-bosom'd night—press close magnetic
    nourishing night!
Night of south winds—night of the large few stars!
Still nodding night—mad naked summer night.
    —Walt Whitman, *Song of Myself*

EARTH'S magnetic field traditionally has been the subject of much speculation. A fundamental understanding of magnetism and the origin and behavior of our planetary magnetic
field is essential before a thorough description can be given of the
magnetic fields of the Sun, other objects in the solar system, and
of such stellar remnants as white dwarf stars and pulsars (rapidly
rotating neutron stars). In this chapter we will explore how
Earth's magnetic field was discovered, what it tells us about the
structure and internal motions of Earth, and how the magnetic
field is generated.

The common magnetic iron-oxide mineral magnetite (Fe_3O_4)
was called lodestone by medieval sailors who used it in com-

passes for navigation. *Lode* is the Middle English term for "way," and lodestone was a stone that could show sailors the way to their destination. Descriptions of this mineral appeared in Greek writings as early as 800 B.C.E. in the period when Homer composed the *Iliad* and the *Odyssey*. It was mined in the Greek province of Magnesia in Thessaly, and the mineral became formally known as magnetite.

Its attractive properties were ascribed to supernatural influences. In recounting observations of the philosopher Thales of Miletus, Aristotle wrote: "Thales, too, as is related, seems to regard the soul as somehow producing motion for he said that the stone has a soul since it moves iron." Diogenes of Apollonia claimed that magnets were inherently dry and fed upon humidity in iron. However, the ancient Greek philosophers did little experimentation with magnets, and the magnet's polarity and orientation properties were not discovered for more than a thousand years.

Although the Chinese had attached a freely rotating magnetite spoon to a smooth board by the second century B.C.E., it was not until 1180 C.E. that the first written reference to the magnetic compass appeared in Europe. Within a few decades it became obvious that the compass needle pointed toward the North Star, Polaris, and the suggestion was made that lodestone derived its properties from that star's influence. In an alternative view, the compass was thought to be affected by mountains of lodestone situated at the poles.

Pierre Pélerin de Maricourt (better known as Petrus Peregrinus) pointed out that lodestone deposits occurred in many localities throughout the globe and it was quite unlikely that compass needles should be attracted only toward such deposits at the poles. In *Epistola de Magnete*, written in 1269, Peregrinus reported his series of experiments with a spherically fashioned lodestone block. After placing an oblong fragment of iron on various places of the lodestone sphere and marking the directions the fragment pointed, he discovered magnetic meridians (lines where the iron fragment had the same orientation), defining the poles of the magnet as the two points where the meridians intersected. Peregrinus was also the first one to note the repulsion of like magnetic poles and the attraction of unlike magnetic poles.

By 1450 it was apparent to European instrument makers that magnetic north and true north (as defined by the geographic pole) were not coincident. This deviation from true north is called magnetic declination, and some sundials from this period were equipped with compass settings appropriately corrected. Drawings of compasses, with needles deviating from true north, also appeared on road maps during the fifteenth century.

Magnetic inclination (the dip angle between the compass needle and the horizontal) was discovered in 1544 by Georg Hartmann. In 1600 William Gilbert, a personal physician of Queen Elizabeth I, published *De Magnete*, in which he described the variation of magnetic inclination over a lodestone sphere. Gilbert concluded that "the Earth globe itself is a great magnet" and reasoned that the geomagnetic field was primarily dipolar. In other words, it resembled a big bar magnet.

Magnetism was thus the first property, aside from the spherical shape of Earth, that was historically ascribed to the planet as a whole. It was not until 1687 that Newton's *Principia* was published and established (among other things) Earth's global gravitation.

Gilbert demystified magnetism by refuting such fabulous contemporary claims as the ability of garlic and diamond to rob magnets of their attractive power. He also scoffed at the widespread notion that lodestone placed beneath the pillow of a sleeping woman would drive an adulteress, but not a virtuous lady, out of her bed.

Gilbert considered Earth's magnetic field to be a constant phenomenon, perhaps modified only by rare geologic catastrophes. In 1635, however, thirty-two years after Gilbert's death, Henry Gellibrand noticed that the value of London's declination in his own time differed from earlier determinations by a greater amount than could be reasonably ascribed to error. This observation implied the existence of a slow, progressive shift of the locations of the geomagnetic poles. This shift is called the secular variation.

The dipolar magnetic field of Earth can be represented as emanating from a uniformly magnetized sphere, as demonstrated by Karl Friedrich Gauss in the nineteenth century, or by a sphere with an internal bar magnet presently offset about 11.7° from the sphere's axis of rotation (figure 4.1). At the high temperatures prevailing inside Earth, however, materials cannot be permanently magnetized, so these models, although descriptive, do not mimic reality.

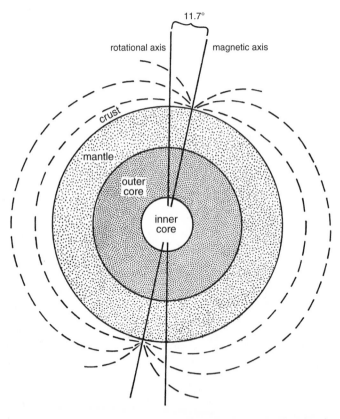

Figure 4.1. Schematic diagram showing Earth's internal structure, rotational axis, and magnetic axis. At present, the magnetic axis is offset by 11.7° from the axis of rotation. Dashed lines represent magnetic lines of force.

At temperatures above about 500°C, the Fe^{2+} and Fe^{3+} ions (charged atoms of iron) in magnetite will be oriented randomly. They vibrate around their mean positions and are incapable of being magnetized by any ambient external magnetic field. A decrease in the temperature of the solid corresponds to a decrease in the amount of heat energy available to the ions. Consequently, the vibrations diminish and the effects of an external magnetic field become more prominent. The temperature at which this happens varies among minerals; it is called the Curie point, after Pierre Curie. Small domains of ions in the grains align themselves parallel with the direction of the field. Once the domains are parallel, the individual ions lack the necessary energy to vibrate out of this ordered arrangement, and each crystal of magnetite, has, in effect, itself become a magnet. Its polarity is exactly parallel to that of the surrounding magnetic field.

The Curie point is exceeded in terrestrial rocks below depths of approximately 25 kilometers, thus proving that Earth cannot be a giant permanent magnet. The temperature is too high; the iron atoms vibrate too rapidly to stay aligned.

Lava flows with melting temperatures in the range of 800°C–1200°C typically acquire the direction of Earth's magnetic field after the flows solidify and cool below the Curie point. The rocks retain a record of the polarity, inclination, and declination of the geomagnetic forces exerted on them at the particular place and time of cooling. Any subsequent changes in the ambient field will not affect the rocks' thermoremanent magnetization (i.e., the magnetization remaining after cooling). Even if the surrounding magnetic field that originally produced it changes, the magnetization of the cooled rocks is an excellent indicator of the paleomagnetic (i.e., former magnetic) conditions. Of course, any reheating above the Curie temperature or changes in the rocks' orientations after the time of primary solidification must by considered.

After solidification, many of the magnetic mineral grains undergo mechanical weathering. They are transported by rivers into the sea, and ocean currents distribute these grains globally. The particles have various settling velocities, but upon finally reaching the ocean bottom, they are affected by the currents only minimally. There they can rotate somewhat freely in the unconsolidated sediment, and at this point the effects of the geomagnetic field again become important. It tends to align the grains parallel to the magnetic lines of force.

As more material settles over the sediment, the sediment gets compacted. Eventually it turns to stone and preserves a permanent record of the magnetic field at the time and place that the grains were deposited.

Sediments generally contain only a small percentage of magnetized grains, and the alignment of the particles with Earth's magnetic field is only approximate. Thus, most sedimentary rocks have an intensity of magnetization about a hundred times weaker than that of volcanic rocks. Between deposition and solidification, these sediments can change chemically and physically. This may alter the magnetization and reduce the reliability of these samples as paleomagnetic indicators.

Several lava flows of the Sicilian volcano Mt. Etna were examined by Raymond Chevallier in 1925. He demonstrated that the rocks' thermoremanent magnetizations were aligned with the geomagnetic field that had been measured by local observatories around the time of the eruption. However, in 1906 Bernard Brunhes discovered that certain lavas from the Massif Central mountain range in France had acquired a remanent magnetization opposite in polarity to the modern geomagnetic field. Other observations followed; reversely magnetized rocks were discovered in locations as diverse as South Africa, Greenland, and Japan. To date, reversely magnetized rocks constitute approximately one-half of all the specimens that have been measured.

These data suggested that Earth's magnetic field has reversed itself in the past, but skeptical researchers pointed out that other possible explanations of the observations of the reversely magnetized rocks had to be discounted (in at least some instances) before spontaneous reversals of Earth's magnetic poles could be accepted as fact. It was conceivable that some rocks, upon cooling through the Curie temperature, would acquire reverse magnetization as magnetic grains oriented themselves opposite to the general magnetic field. The skeptics also asked whether a rock magnetized parallel to the direction of Earth's magnetic field could spontaneously reverse its own magnetization.

Field observations of lava flows that have cut across and baked sedimentary rocks go far to show that spontaneous reversals of the rocks' magnetism are unlikely to be significant. A preexisting sediment, baked by lava or an igneous intrusion, will rapidly increase in temperature and exceed the Curie point. Upon cooling, the magnetic grains in the sediment will orient themselves to the direction of the ambient magnetic field. In every case, the baked sediments and the crosscutting lavas or intrusions have the same polarity, whether or not this polarity is parallel, or antiparallel, to the present geomagnetic field. It seems quite unlikely that the sediments and the igneous rocks, often vastly different in bulk composition, should possess the same spontaneous reversal characteristics. Changes in the direction of Earth's magnetic field are more likely.

In the early 1950s French physicist Louis Néel envisioned possible ways in which a rock might spontaneously acquire a magnetization opposite that of an external magnetic field. One way involves a hypothetical substance that is composed of two different magnetic minerals, one possessing a low intensity of magnetization and a high Curie temperature and the other a high intensity of magnetization and a low Curie temperature. After initial cooling of this substance from a melt, the mineral with the higher Curie point becomes oriented in the direction of the external magnetic field. When the temperature drops to the lower Curie point, the second mineral is subjected to two magnetic influences: the general field and the already magnetized grains of the first mineral. The magnetic field of the second mineral (with its higher intensity of magnetization) may become oppositely directed to that of the first and in doing so give the entire rock a net reversed polarity.

A pumice from Mt. Haruna, an extinct volcano in Japan, was examined by Takesi Nagata and coworkers in 1952 and found to be reversely magnetized while still in place. In the laboratory it was shown that this rock had acquired a net magnetization antiparallel to the field in the way Néel described it might. However, only a very few igneous rocks that show magnetic reversal in the field show this reversal property in the laboratory.

The similarities in magnetic polarity of baked sediments and crosscutting lavas, the fact that few igneous rocks found in situ to be reversely magnetized behave in the laboratory according to Néel's mechanisms, and the simultaneous occurrence all over the earth of reversely magnetized rocks from the Pennsylvanian and Permian periods of geological history (approximately three hundred million years ago) all provide impressive evidence for spontaneous reversals of Earth's magnetic poles.

One of the first persons to study the history of changes in geomagnetic polarity was Motonori Matuyama. In 1929 he published a paper discussing his investigations of 139 basalt specimens from thirty-eight different localities in Japan, Korea, and Manchuria. He found that the rocks' magnetic polarities fell into two distinct groups: one approximately parallel to the present ambient field in Japan, the other antiparallel to the present field. By correlating approximately determined ages of individual rocks with their magnetic polarities, Matuyama was able to demonstrate that the reversely magnetized basalts were all formed no more recently than the Pleistocene while the normally magnetized basalts were of more recent Quaternary age. He concluded that "the Earth's magnetic field in the present area has changed even to opposite direction in . . . (the) Miocene and also in Quaternary periods." The current estimate of the time of the latter geomagnetic reversal is seven hundred thousand years before present.

Later polarity measurements (such as those in 1963 by Ian McDougall and Don Tarling on Hawaiian lavas and in 1964 by Allan Cox and coworkers on volcanics from Idaho, Alaska, California, Hawaii, and New Mexico) went far in proving the worldwide simultaneity of reversals in Earth's magnetic poles. By 1969, Cox was able to assemble a table of all the geomagnetic epochs

Eon	Era	Period	Epoch	Time since Beginning (millions of years)
Phanerozoic	Cenozoic	Quaternary	Holocene	0.01
			Pleistocene	1.6
		Tertiary	Pliocene	5.2
			Miocene	23.3
			Oligocene	35.4
			Eocene	56.5
			Paleocene	65.0
	Mesozoic	Cretaceous		146
		Jurassic		205
		Triassic		251
	Paleozoic	Permian		290
		Carboniferous	Pennsylvanian	323
			Mississippian	362
		Devonian		408
		Silurian		439
		Ordovician		510
		Cambrian		570
Precambrian	Proterozoic			2500
	Archean			3800
	Hadean			4560

of the last four and a half million years, including events of quite short duration (figure 4.2).

The timescale could not be extended farther back, because the potassium-argon technique used in dating the rocks was insufficiently precise to enable the ages of older rocks to be accurately determined. A 5 percent dating error in a five-million-year-old sample is 250 thousand years, longer than the entire duration of some of the polarity-reversing events (a few of which lasted only ten thousand years.) Although all of the magnetic epochs for the past four and a half million years have already been discovered,

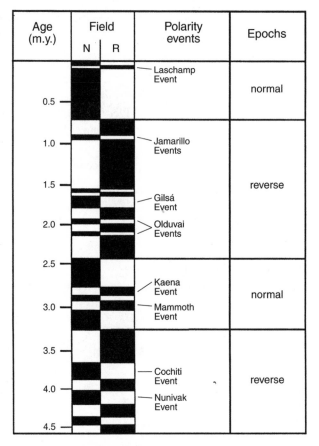

Age (m.y.)	Field		Polarity events	Epochs
	N	R		
0.5			— Laschamp Event	normal
1.0			⌐ Jamarillo Events	
1.5				reverse
2.0			⌐ Gilsá Event ⌐ Olduvai Events	
2.5				
3.0			⌐ Kaena Event — Mammoth Event	normal
3.5				
4.0			— Cochiti Event — Nunivak Event	reverse
4.5				

Figure 4.2. Column representing magnetic reversal events during the past 4.5 million years. Periods when the magnetic field was most often aligned in the same direction as its present alignment are called normal epochs; periods when the field was most often oppositely aligned are called reverse epochs. Short-period reversals are known as polarity events.

there are a few gaps in the data covering intervals longer than one hundred thousand years, indicating that there may be short-period events that remain undetected.

A continuous sequence of geomagnetic polarity determinations based on lavas alone is not possible, because volcanoes erupt intermittently. Deep-sea sediment cores must also be examined,

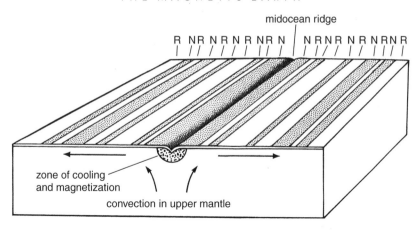

Figure 4.3. Basaltic lava flows erupting at midocean ridges acquire the magnetic orientation of Earth's field as they cool through the Curie points of their magnetic minerals. With time, convection currents in the upper mantle carry the basalt flows away from the ridge, and new basalts erupt. In this way, the ocean floor develops magnetic stripes in its basalts parallel to the ridge; the stripes are symmetrically distributed on both sides of the ridge.

both to independently verify the polarity reversals deduced from the lavas and to extend geomagnetic history further.

One sophisticated instrument, the proton precession magnetometer (routinely towed behind oceanographic vessels) has been used since the 1950s to measure the intensity of magnetization of the ocean floor. When enough data had been gathered for plotting on a map, the contours of equal magnetic intensity seemed to be unrelated to ocean floor topography. Instead, a pattern of linear features was apparent, stripes of high magnetic intensity alternating with stripes of low magnetic intensity (figure 4.3). These "magnetic stripes" varied from about 2 to 80 kilometers in width, all being several thousand kilometers in length. At fracture zones, the stripes abruptly ended and were shifted from the next set by distances ranging up to more than a thousand kilometers. This represented a new global phenomenon, one that defied satisfactory explanation until 1963.

By 1963 the magnetic stripes had been observed over the North Atlantic, the Antarctic, and the Indian Oceans. Fred Vine and Drummond Matthews suggested that about 50 percent of the oceanic crust was reversely magnetized. Using the current ideas on seafloor spreading (advocated by Robert Dietz and Harry Hess) and periodic geomagnetic polarity reversals (compiled by Allan Cox), Vine and Matthews proposed that while lava is extruded at the center of oceanic ridges it acquires the direction of the general geomagnetic field in the neighborhood, whether this field is normal or reversed. As seafloor spreading proceeds, alternate blocks of normally and reversely magnetized basalt drift away symmetrically from the ridge axis.

Stripes of high magnetic intensity arise from the oceanic basalt that is normally magnetized, for the basalt's magnetization and the field surrounding of the present era have the same direction. On the other hand, when a basalt layer has been reversely magnetized, a distant magnetometer (such as one towed across the ocean surface) will measure a combination of the general field and the oppositely polarized, older rocks. This shows up as an overall low-intensity magnetic "stripe." By correlating the magnetic stripes flanking the oceanic ridges with the paleomagnetic and radiometric data from the hundreds of lava measurements, the spreading rates of the lithospheric plates can be estimated, and the history of geomagnetic polarities can be extended much further into the past.

The Vine and Matthews interpretation of the magnetic stripes led directly to the concept of plate tectonics—the idea that Earth's crust and upper mantle are divided into about a dozen major "plates" that drift across Earth's surface, floating on a spherical shell of less-dense material. Collisions between plates cause earthquakes, volcanism, and mountain building. Sites of ancient mountains such as the Urals and Appalachians mark the boundaries of ancient plate collisions.

By 1968 the magnetic stripes had been used to construct a geomagnetic timescale stretching back into the Cretaceous. Although deep-sea drilling in the South Atlantic subsequently confirmed this scale, several detailed revisions have been necessary. The polarity of the geomagnetic field has changed sporadically: the frequency of geomagnetic reversals was high in the Upper Tertiary (Pliocene and Miocene) and somewhat lower in the Early Tertiary (Paleocene) and Upper Cretaceous. More recent constructions of timescales, based in part on geochronometry of ocean sediment cores, indicate that the reversal frequency was very low throughout the Lower Cretaceous and higher in the Upper Jurassic.

Geophysicists hoped to find rocks or sediments that acquired their remanent magnetization during a transition interval from one polarity to the other in order to see how the geomagnetic field behaved during these periods. However, such evidence is rare. Lavas are extruded inconstantly, and the sedimentation rate on the ocean floor is quite slow. Polarity-reversing events, on the other hand, may be as short-lived as a thousand years. Nevertheless, geophysicists have discovered a few cases where polarity transitions occurred over successive lava flows in South Africa, Oregon, and southeastern Australia.

Reversals of Earth's magnetic field have occurred at least since the Cambrian, and any explanation of these reversals must take this into account. In 1929, when Matuyama first proposed a polarity reversal, geophysicists accepted his ideas as possible. As Cox noted in 1973, "[at that time] geophysicists still did not have a sensible theory for the origin of the geomagnetic field. And so long as geophysicists did not know why the field was pointing north, there was no reason to reject the idea that it might at another time have pointed south." This reasoning assumes that any geomagnetic reversals that may have occurred arose from purely internal processes. Geophysicists believed that once the source of geomagnetism was discovered, the reason for polarity reversals would soon become apparent. The fact that the intensity of Earth's field progressively diminishes before it reverses strongly supports an internal origin for the field.

Bill Glass and Bruce C. Heezen suggested in 1967, however, that impacts of giant meteorites with Earth's surface caused geomagnetic reversals. This hypothesis was based upon the approximate simultaneity of the last major polarity reversal and the formation of the Australasian tektites seven hundred thousand years ago. Although the tektites were indeed produced by impacts (as we will see later), this similarity in age is only coincidental. The ages of large impact craters do not correlate well with the dates of known polarity reversals.

Instead, the geomagnetic reversals are probably due to some internal, rather than external, mechanism. Evidence to support this arises from a number of sources. The significantly decreased velocity of P-waves (waves of seismic compression) and the complete absence of S-waves (shear waves) between depths of about 3,000 to 5,000 kilometers inside the earth indicate the existence of a fluid outer core. (Other seismic data also indicate the presence of a solid inner core.)

What is the composition of the core? Because most iron meteorites are thought to have resulted from the fragmentation of a cosmic body structurally differentiated into a metallic core and rocky mantle, geophysicists assume that iron is greatly concentrated in Earth's interior. The fact that Earth's density is about 5.5 grams per cubic centimeter, whereas the density of surface rocks is generally less than 3 grams per cubic centimeter, implies the existence of large amounts of much denser material deep inside Earth. The element iron, with a density of about 7.9 grams per cubic centimeter (at 1 atmosphere pressure), seems a likely candidate for this core material. A fluid core of iron, replete with convective and/or rotational motion, represents a reasonable model of the inside of the planet and is very probably responsible for the generation and behavior of Earth's magnetic field.

The iron in Earth's core originated in the planetesimals that accreted to form the planet. The planetesimals themselves may have been chondritic, consisting of a homogeneous mixture of metal, sulfide, and silicate. Alternatively, the planetesimals could have been former chondritic bodies that subsequently melted and differentiated into objects with silicate mantles and metallic iron cores.

Since many paleomagnetic measurements have demonstrated that Earth's magnetic field has existed throughout geologic time, never significantly decreasing in intensity except during transition intervals, the field cannot be simply a leftover from the past. Electrical currents in a finite electrically conducting body will eventually decay. It has been estimated that the slowest decay for a spherical object the size of Earth would significantly decrease the magnetic field strength within about one hundred thousand years. Earth is approximately forty-six thousand times older than this. Thus, a mechanism must be found that allows the geomagnetic field to be continuously generated.

Sir Joseph Larmor suggested, as early as 1919, that the magnetic field of the Sun might be powered by an internal generator called a dynamo. Later workers applied an analogous mechanism to the magnetic field of Earth. Dynamo theories, whether they are applied to planets or stars, all derive their foundations from a branch of science known as magnetohydrodynamics. This science is concerned with the motion of an electrically conducting fluid in the presence of a magnetic field. The motion of an electrically conductive fluid produces electric currents, and these alter the magnetic field. The flow of these currents in the field at the same time produces physical forces that alter the motion. This magnetic field/fluid motion interaction forms the basis for the various dynamo models of Earth's core.

The 1,000-year period apparently required for the transition intervals of geomagnetic reversals agrees with the length of time theoretical magnetohydrodynamic processes take to work in Earth's core. The long-term variation also indicates that the liquid outer core is in motion. Explanations of Earth's magnetic field and the reversals of its poles must account for the long-term variation and the lengths of transition periods. Strong support for a dynamo model of Earth's core arises from computer simulations of certain complex arrangements of dynamo components. The simulated magnetic field generated by the computer undergoes spontaneous polarity reversals in a pattern quite similar to geo-

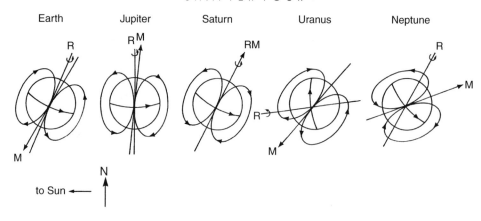

Figure 4.4. The tilts, rotational axes, and magnetic axes of the outer planets compared to those of Earth. In Saturn, the rotational and magnetic axes are essentially colinear, but these axes are significantly offset in Uranus (59°) and Neptune (47°). The magnetic axes (M) point toward the north end of the dipolar fields generated within the planets.

magnetic reversals. Geophysicists and space physicists now believe that the Sun, Earth, and giant planets all generate their own magnetic fields by internal dynamos (figure 4.4).

Just as the study of geomagnetism led to the discovery of plate tectonics and elucidation of the long history of Earth, geologists in the preceding two centuries had laid the groundwork for the decipherment of Earth's more recent history. They discovered that Earth had once been covered by kilometer-thick sheets of ice. Geophysicists came to realize that the ice ages were caused by the gravitational influences of other bodies. It brought home the fact that Earth is not an isolated body but is instead a fundamental part of a complex solar system. The story of the discovery of the ice ages is another example of how science is actually done. Researchers make mistakes, puzzle over contradictory evidence, argue with one another, and sometimes change their minds. This story is told in the next chapter.

V

ICE AGES

O

And now there came both mist and snow,
And it grew wondrous cold:
And ice, mast-high, came floating by,
As green as emerald.
The ice was here, the ice was there,

The ice was all around:
It cracked and growled, and roared and howled,
Like noises in a swound!

—Samuel Taylor Coleridge, *The Rime of the Ancient Mariner*

Nothing could have been more obvious to eighteenth-century European naturalists than the geological evidence for the biblical Deluge. As explained in Genesis 7:19, "the waters prevailed exceedingly upon the earth; and all the high hills, that were under the whole heaven, were covered." Large boulders were scattered across northern Europe; some were huge and weighed thousands of tons. They lay in fields, perched on hillsides, and on the floors of U-shaped valleys. Many bore no resemblance to local rocks and were thus called "erratic"; they must have traveled hundreds of kilometers from their source. The naturalists reasoned that the boulders had been transported by powerful torrents unleashed when "the fountains of the great deep broke up and the windows of heaven were opened."

In addition to the erratic boulders, there were chaotic deposits of clay, silt, gravel, and sand in the mountains and valleys. Many workers ascribed these so-called drift deposits to the Flood, but the problem remained as to how they could have been emplaced at high elevations. Sea level must have risen over 1,500 meters. Although higher seas were consistent with the biblical account, naturalists were puzzled about where all the water could have gone. Some postulated that it drained deep into the earth through uncharted caverns. Others tried to avoid the problem by suggesting that sea level had not changed at all; the high-elevation sedimentary deposits had instead resulted from mile-high waves generated by a hypothetical rapid wobbling of Earth on its axis or by a giant comet grazing Earth's surface.

Another curious phenomenon was the occurrence in drift deposits of polished rocks with numerous scratches. Flood aficionados postulated that pebbles suspended in turbulent waters had banged into larger rocks and gouged them.

The Swiss physician and naturalist Johann Jacob Scheuchzer traveled the countryside in the late eighteenth and early nineteenth centuries collecting fossils and ascribing their origin to Noah's Flood. In 1708, while walking with a companion through the small town of Altdorf, Scheuchzer stopped at the foot of the municipal gallows and poked at a slab of dark marble containing eight vertebrae. In his 1726 report, Scheuchzer described the fossils as consisting of "numerous parts of the human head—genuine residuaries distinguishable from any other class of animals: the brain case, the frontal bone, the sincipital and occipital bones, the orbits, the base of the brain and the remains of the medulla oblongata." He also described a more complete skeleton that had been unearthed from a nearby quarry. From its proportions, he estimated that the "total stature of the man approximated [his] own: 58 and one-half Parisian thumbs." He named the new fossil species *Homo diluvii testis*—"the man who witnessed the Flood."

In one account of the bones, Scheuchzer appended a couplet in hopes that his contemporaries would draw the appropriate moral lesson:

> Betrübes Beingerüst von einem altem Sünder
> Erweiche, Steine, das Herz der neuen Bosheitskinder!

In an unpoetic translation, this couplet would mean:

> Afflicted skeleton of an ancient sinner
> Soften, Stone, the hearts of the modern children of malice!
> (i.e., the present generation of spiteful people)

The true nature of *Homo diluvii testis* was not uncovered until 1825, when the great French anatomist Baron Georges Cuvier described the bones as belonging to an extinct giant salamander. Cuvier remarked with some astonishment that "[n]othing less than total blindness on the scientific level can explain how a man of Scheuchzer's rank, a man who was a physician and must have seen human skeletons, could embrace such a gross self-deception."

Some geologists were also skeptical that powerful currents could move house-size boulders great distances and then deposit them at high elevations. In 1833 the renowned British scientist Charles Lyell (regarded by historians of science as the father of geology) published the first edition of his classic work, *Principles of Geology*. In the book, Lyell suggested that the erratic boulders had been entombed in icebergs and set adrift in the Flood. This seemed consistent with recent reports from polar explorers who had observed icebergs calving off glaciers at the point where the glaciers entered the sea. Lyell surmised that the icebergs would melt as they drifted to lower latitudes and their sediment loads would end up on the ocean floor. Lyell's drift theory received support in 1839 after Charles Darwin published an account of his voyage on the H.M.S. *Beagle* that included a report of icebergs in the southern ocean that contained boulders.

Lyell also suggested that the chaotic drift deposits of sand, gravel, and silt in the mountains had been derived from sediment-laden ice rafts. He proposed that earthquake-induced avalanches had dammed mountain streams and formed large lakes upon which ice rafts could drift.

Another distinguished geologist who pondered the origin of erratic boulders and drift deposits was Lyell's former teacher, the Reverend William Buckland, professor of mineralogy and geology at Oxford. Buckland was a man of abiding faith. He felt that the purpose of geology was to "confirm the evidences of natural religion; and to show that the facts developed by it are consistent with the accounts of the creation and deluge recorded in the Mosaic writings." In 1823 Buckland published a massive volume describing the geological evidence for the Flood: *Reliquiae Diluvianae; or, Observations on the Organic Remains Contained in Caves, Fissures, and Diluvial Gravel, and on Other Geological Phenomena, Attesting the Action of an Universal Deluge.* For this monumental work, Buckland was awarded the Copley Medal of the Royal Society.

Buckland's inclination was to ascribe the erratic boulders, scratched and polished rocks, and drift deposits to the Flood, but despite his monograph and his medal, his field observations left him unconvinced. He examined Lyell's iceberg model but found it unsatisfactory. What he needed was a different theory, one that had scientific evidence supporting it.

The theory that Buckland eventually embraced was that large glaciers—ice sheets—had once covered northern Europe. The glacial theory had actually been advanced in the eighteenth century by Swiss villagers who had recognized that the native bedrock of the Jura Mountains (which occur along Switzerland's border with France) was mainly limestone whereas many erratic boulders on the southeastern flanks of the mountains were granite. They inferred that the granite boulders had been transported by enormous glaciers that had extended to the Jura Mountains from the Alps, more than 100 kilometers to the southeast.

While tracking chamois (a fleet-footed animal sometimes called the goat-antelope) in the Alps, a hunter named Jean-Pierre Perraudin had occasion to examine the rocks along his way. In the valleys, he observed scratched and polished rocks that had identical markings to the rocks surrounding the mountain glaciers. By 1815 he had concluded that if glaciers could scratch the rocks in the mountains, they must have once done so in the valleys. He contacted a naturalist named Jean de Charpentier, director of the salt mines at Bex, Switzerland. Although interested in Perraudin's observations, Charpentier found the glacial hypothesis "so extraordinary and even so extravagant that [he] considered it not worth examining or even considering."

Despite such setbacks, Perraudin was able to recruit one supporter, a civil engineer named Ignace Venetz. Venetz began studying glacial moraines and the poorly sorted rock fragments, sand, and silt composing them. These deposits, now known as glacial till, were once part of a glacier's sediment load and deposited at the terminus of a glacier as the ice melted and sublimated. Venetz recognized that the drift deposits strewn across northern Europe resembled glacial moraines and reasoned that the erratic boulders had been dumped where they lay by receding glaciers. He concluded that glaciers had once covered all of Switzerland and parts of other European countries. Venetz presented his ideas at scholarly meetings in the 1820s but encountered only rejection.

The pattern of the lone scholar advocating extensive glaciation in past times to skeptical scientists was to be repeated. Also to be repeated, however, was the pattern of dedicated scientists changing their minds after being confronted with solid evidence of such glaciation.

After hearing Venetz speak in 1829 at the annual meeting of the Society of the Hospice of the Great St. Bernard, Charpentier mulled over the glacial theory he had rejected fourteen years earlier. Venetz's more extensive data impressed Charpentier and he soon embraced the glacial model. Intent on refining the model, Charpentier committed himself to several years of field studies. He traced erratic boulders in the Jura Mountains back to their source in Alpine valleys. He deduced that the striated rocks were scratched by pebbles embedded in glacial ice being dragged across them. Charpentier presented his ideas in 1835 at a meeting in Lucerne but, like Perraudin and Venetz before him, was met with a cold reception.

In attendance at the meeting was the eminent naturalist Louis Agassiz, a fossil-fish expert and longtime friend of Charpentier. Agassiz rejected the glacial theory and set himself the task of showing his friend the error of his ways. The two visited several glaciers and moraines, and Agassiz had Charpentier lower him with a rope into open crevasses so that he could closely examine the ice and its sediment load. On one occasion, Charpentier unknowingly lowered Agassiz into near-freezing meltwater at the base of the ice. This inadvertent baptism helped convert Agassiz to the glacial theory when he realized that the meltwater could lubricate the base of glaciers and facilitate their movement.

Agassiz later borrowed the notes of biologist Karl Schimper, who had studied erratic boulders in Bavaria. Schimper had concluded that thick ice had once covered much of Europe, North America, and northern Asia. In honor of Galileo's 273rd birthday in February 1837, Schimper composed a poem titled *Die Eiszeit* (The Ice Age) and thereby introduced a now common term. Agassiz adopted the term and handed out copies of Schimper's poem at his lectures.

In July 1837 Agassiz used his presidential address to the Swiss Society of Natural Sciences at Neuchâtel to present the glacial theory. He summarized the work of Charpentier and Schimper and postulated a generally colder past climate in the Northern Hemisphere. He argued that these conditions caused widespread faunal extinctions including that of Siberian mammoths. Many attendees had expected to hear about Agassiz's studies of fossil fish and were surprised to find him present such a preposterous hypothesis. Friends urged Agassiz to go back to fossil fish, but he was not to be dissuaded. In 1840 he published his classic monograph, *Studies on Glaciers*, in which he showed that Alpine glaciers had grown and shrunk repeatedly in historic times. It did not seem far-fetched to suppose that they had once been far more extensive than at present.

While preparing his monograph, Agassiz was visited in Neuchâtel by William Buckland and his wife. Agassiz showed Buck-land the erratics in the Jura Mountains and the glaciers in the Alps. Buckland warmed to the glacial theory but did not become fully convinced until 1840, after he and Agassiz had examined moraines, erratic boulders, and striated rocks in northern England and Scotland. While in Glasgow, Agassiz presented his views at a meeting of the British Association for the Advancement of Science and was pointedly attacked by Charles Lyell. However, Buckland had recently become an energetic proponent of the glacial model and took it upon himself to persuade Lyell.

By November 1840 Lyell had been won over and presented a paper, "On the Geological Evidence of the Former Existence of Glaciers in Forfarshire," to a meeting of the Geological Society in London. Agassiz attended the meeting and lectured on "Glaciers and the Evidence of Their Having Once Existed in Scotland, Ireland, and England." Buckland's talk was titled "Evidence of Glaciers in Scotland and the North of England." Although the presentations by the three prominent geologists met with some skepticism, several converts were made. The following year, British geologist Edward Forbes wrote to Agassiz: "You have made all the geologists glacier-mad here, and they are turning Great Britain into an ice house."

In 1846 Agassiz sailed to the New World and immediately discovered evidence for past glaciation in Nova Scotia. Upon arriving in Boston, Agassiz was pleased to find that a number of American geologists had already accepted the glacial theory. He was offered a professorship at Harvard and remained in America the rest of his life.

Many geologists embraced glacial theory after an 1852 expedition to Greenland showed that this large landmass was almost entirely covered by an ice sheet. Ten years later, Scottish geologist Thomas F. Jamieson described the effects of a dam burst in the Scottish Highlands that had caused the rapid emptying of several water reservoirs. Although the rushing water contained pebbles, it did not produce striated rocks or poorly sorted sedimentary deposits resembling those associated with glaciers. This report served as the coup de grâce of Flood geology for most British scientists, and glacial theory became the ruling paradigm.

After this, many geologists worked to fill in details of the Ice Age. They investigated its duration, the extent of ice sheets, and whether there had been multiple episodes of glaciation. And physicists, convinced by their geologist colleagues that there had actually been an Ice Age, began wondering about its cause.

Geologists exploring the formation of glaciers discovered that glacial ice forms beneath snowpacks that are more than 30 meters high. The enormous pressure of the overlying snow causes the ice at the sharp points of underlying (basal) snowflakes to melt, reducing flake porosity. Eventually, the snowflakes transform into millimeter-size ellipsoidal granules of ice. Continued accumulation of snow increases the pressure, drives out the air, and cements the loosely packed granules together. Daily temperature fluctuations tend to produce some meltwater near the base of the snowpack that then seeps through pore spaces between ice granules. Freezing of the meltwater fills the pore spaces and helps form compact glacial ice. Where snowpacks exceed about 60 meters in height, basal ice can deform plastically and flow downhill (figure 5.1). Basal ice can also melt and refreeze, causing the glacier to slip over the bedrock.

To measure glacial flow, Swiss geologists placed boulders in straight lines across mountain glaciers. Within a few years, they observed that the lines were no longer straight: the boulders nearest the center (where the glacier was thicker) had traveled the farthest downslope; those nearest the valley walls had traveled the shortest distance.

Beneath a glacier, meltwater seeps into rock fractures, refreezes, and expands. The pressure of expansion causes bedrock fragments to wedge loose, break off, and become incorporated into the ice. The rock fragments within the ice grind against the bedrock, gouging it out and carving valley walls with U-shaped cross sections (figure 5.2). Sand-size particles within glacial ice polish and scratch underlying bedrock. At the glacier terminus, where the ice melts as fast as the glacier flows downslope, rocks

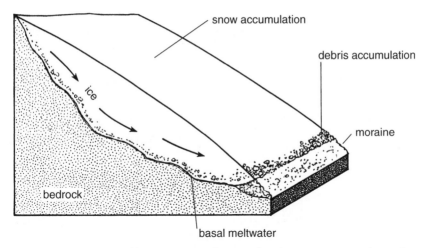

Figure 5.1. The weight of large amounts of accumulated snow compacts the under-lying snow, forming glacial ice. Meltwater may form at the base of the glacier, aiding its downhill slide. As it moves downslope, the glacier gouges out chunks of bedrock and transports the rocky debris. Where the front of the glacier is melting at the same rate that the glacier is flowing forward, debris is deposited in a mound called a moraine.

of all sizes moving within the ice come to a halt and are deposited into mounds known as terminal moraines in front of the glacier. As glaciers recede, terminal moraines mark their maximum advance.

Geologists were able to trace terminal moraines across North America from New England to the state of Washington (figure 5.3). Similar maps of terminal moraines were made for Europe, Asia, Australia, New Zealand, and South America. By 1875 geologists were able to use these maps to determine that ice sheets had once covered three times as much of Earth's surface as they do today. There had been only a paltry increase in ice cover in the Southern Hemisphere, but in the north the ice had once covered thirteen times more land than at present.

After reading Agassiz's monograph on glacial theory, Scottish geologist Charles Maclaren had realized as early as 1841 that with so much water locked up in glacial ice, there would have to

Figure 5.2. Glaciers erode underlying rocks to form U-shaped valleys such as this one in Rocky Mountain National Park in Colorado.

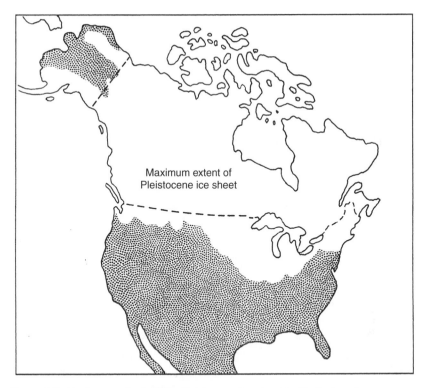

Figure 5.3. Maximum extent of the Pleistocene ice sheet in North America.

have been a significant decrease in sea level. Modern estimates put this decrease at about 100 meters. This idea was substantiated in the 1860s by the discovery of submerged wave-cut cliffs at several sites in northern Europe.

The lower level of the Bering Sea during the Ice Age exposed dry land on the continental shelf along the Bering Strait connecting Asia and North America. Most archeologists came to believe that about thirty thousand years ago (the date is controversial) Stone Age hunters followed migrating herds of game animals across the strait. These hunters and their families became the first Americans.

Similarly, dry land appeared between the Japanese islands, the Korean Peninsula, and eastern Siberia, allowing a different group of Stone Age hunters to populate Japan.

In contrast to the lower level of the sea during the Ice Age, immediately after the glaciers retreated, the sea level in northern Europe relative to the land surface was appreciably higher than it is today. Scandinavian geologists identified marine shells in the mountains at elevations exceeding 300 meters above present sea level. Some workers reported shorelines of ancient beaches at high elevations. In 1865 Thomas Jamieson suggested that the great weight of the ice sheet had depressed the land, resulting in local inundation when the ice melted. He postulated that beneath Earth's rigid crust was a layer containing molten rock that the crust displaced when weighed down by the glaciers. This layer, now known as the asthenosphere, was not confirmed by geophysicists until the early twentieth century. With removal of the icy burden, the land began to rise at a rate of about 2 centimeters per year toward its former elevation.

As the ice retreated, streams of meltwater deposited great quantities of silt in front of the glaciers. This loose, fine-grained material was not anchored to the ground by vegetation and was readily picked up by high winds and blown great distances. The rich farmland of the American Midwest is composed largely of this windblown silt. The silt is known today by the old German term *loess* (roughly pronounced "luss"). Loess covers more than 2.5 million square kilometers of North America, Asia, and Europe, in some places to depths exceeding 3 meters. Its origin as a windblown deposit was recognized in 1870 by German geologist Ferdinand von Richthofen, who pointed out that loess lacked marine fossils and occurred on hillsides at elevations above 2,400 meters. It was thus unlikely to have been deposited by water.

The question of whether or not there had been multiple Ice Ages was broached by geologist Edouard Collomb in 1847. He mapped the Vosges Mountains in the Alsace region of France and found two distinct layers of glacial till separated by well-sorted sediments deposited by streams. Other geologists reported outcrops consisting of separate layers of till in Switzerland, Wales, and Scotland. In 1873 American geologist Amos H. Worthen described humus soil between layers of glacial till in Illinois.

Humus-rich soils form in warm climates from abundant vegetation, indicating that there had been a significant interglacial period when temperatures were moderate. Other American geologists reported the remains of a forest in sedimentary layers separated by till. In his 1874 book *The Great Ice Age*, Scottish geologist James Geikie showed that in some areas there were six distinct glacial epochs separated by warmer interglacial periods. It was clear that there had been multiple episodes of glaciation during the Ice Age.

Theories abounded as to what process may have been responsible for the advance and retreat of glaciers. Some workers suggested a change in the amount of sunlight received by Earth, perhaps due to fluctuations in solar output or to absorption by interplanetary dust grains. Reasoning that the atmosphere is colder at higher elevations, Charles Lyell proposed that tectonic forces in Earth's crust had thrust the land upward. This hypothesis was rejected by James Geikie, who could not believe that such large crustal movements could have occurred in so short a time.

A more promising idea was that a series of cataclysmic volcanic eruptions propelled enough dust into the upper atmosphere to reflect appreciable sunlight into space. This idea received support from the August 1883 eruption of Krakatoa, a large volcano located in the strait between Java and Sumatra. It caused an enormous explosion, equivalent to 100 megatons of TNT; huge quantities of dust were injected into the stratosphere and dispersed around the globe. Until the dust settled, there were unusually red sunsets; mean global temperatures remained a few degrees below normal for several years. However, geologists could find no evidence of anomalously high concentrations of volcanic ash associated with glacial advance. It seemed clear that a different Ice Age trigger was required.

In 1842 the French mathematician Joseph A. Adhémar suggested that ice ages could have resulted from the precession of Earth's rotational axis and the position of Earth in its orbit. The gravitational pulls of the Moon and Sun on Earth's equatorial bulge cause the rotational axis to wobble or precess with a period of 25,800 years. The axis presently points to within 1° of Polaris in the constellation Ursa Minor; 12,000 years from now, Vega in Lyra will be the polar star. Adhémar reasoned that an ice age could start if winter began when Earth was at the farthest point from the Sun in its orbit (i.e., at aphelion).

Adhémar's work was extended in 1864 by James Croll, a Scottish polymath who had worked as a mechanic, carpenter, tea shop proprietor, innkeeper and janitor. While employed as a janitor at the Andersonian College and Museum in Glasgow, Croll took advantage of the scientific library. He familiarized himself with the work of French astronomer Urbain Leverrier, who had calculated variations in the eccentricity of Earth's orbit. Croll found that such variations were cyclical over periods of tens of thousands of years and thought that they might be responsible for the ice ages. He surmised that several conditions had to be met before an ice age could commence in the Northern Hemisphere: (1) it had to begin in the wintertime, when the Sun was low on the horizon and a buildup of snow and ice could occur; (2) it had to begin during epochs of high orbital eccentricity, when Earth could be quite distant from the Sun, near aphelion; and (3) it had to begin when the precession of the rotational axis caused Earth to be at aphelion at the time of the winter solstice (i.e., at the beginning of winter). This situation differs significantly from the present one, in which Earth's orbit is nearly circular and northern winter begins near perihelion (i.e., when Earth's orbit takes it closest to the Sun).

For his work, Croll was offered a position at the Geological Survey of Scotland, where he began to consider how variations in the tilt of Earth's axis relative to the ecliptic plane may also have contributed to the start of an ice age. In 1875 he published his findings in a book, *Climate and Time*, where he predicted that during periods of high orbital eccentricity, ice ages should occur alternately in the Northern and Southern Hemispheres every eleven thousand years. One year later, Croll was elected a Fellow of the Royal Society of London. Despite such accolades, many geologists, particularly those in America, rejected Croll's astronomical theory because of inconsistencies with the (highly uncertain) dates derived for glacial till. Skepticism mounted, and by the time Croll died in 1890, most geologists had dismissed his theory.

In 1904 the German mathematician Ludwig Pilgrim calculated variations in Earth's orbital eccentricity, precession, and tilt over the last million years. These data proved invaluable a few years later to Serbian theoretician Milutin Milankovitch, who took it upon himself to extend the astronomical theory. Milankovitch, professor of applied mathematics at the University of Belgrade, was able to show that the orbital variations considered by Pilgrim were responsible for changes in the amount of sunlight reaching Earth at different seasons and in different locations. Milankovitch calculated that these changes were sufficient to account for both glacial and interglacial epochs.

After reading Milankovitch's 1920 book *Mathematical Theory of Heat Phenomena Produced by Solar Radiation*, German climatologist Wladimir Köppen realized that it was the amount of heat received in the summer (not the winter) that controlled glacial melting. A decrease in summer sunlight could lead to glacial advance if the glaciers melted less in the summer than they advanced in winter. At the behest of Köppen and German geologist Alfred Wegener (the early proponent of continental drift), Milankovitch made more detailed calculations. He explored the variations over the last 650,000 years in the amount of summer sunlight received at eight different latitudes from 5°N to 75°N. His findings were published in 1930 as *Mathematical Climatology and the Astronomical Theory of Climate Changes*.

Milankovitch found that locations at high latitudes are most affected by the 41,000-year cyclical variation in the inclination of Earth's rotational axis; decreased inclination results in decreased summer sunlight. In contrast, low latitudes are most affected by a 22,000-year variation in the precession cycle; for example, temperatures are appreciably lower when Earth is at aphelion at the winter solstice. Milankovitch next considered the snow line—the elevation mark above which snow cover in a given location persists throughout the year. At the poles, the snow line is at sea level; at the equator, it is at the tops of the highest mountains. Milankovitch determined how changes in summer sunlight affect the elevation of the snow line; a decrease in snow-line elevation corresponds to increased glaciation. Unlike Croll, Milankovitch determined that there would be simultaneous ice ages in the Northern and Southern Hemispheres, although the severity of the glaciation need not be the same.

After these studies were published, geologists took a renewed interest in the timing of the ice ages. Tills and interglacial deposits were identified in North America and Europe, but precise dating was difficult. However, it was clear that cyclical variation had occurred. For example, in 1969 geologist Rhodes W. Fairbridge described nineteen evenly spaced parallel sand ridges representing a series of ancient shorelines along the southern coast of Australia. The shorelines resulted from periodic fluctuations in sea level, which in turn were caused by the advance and retreat of massive ice sheets.

In the 1960s geophysicists and geochemists collaborated in studying geomagnetic reversals. They measured the direction of magnetization of lava flows and dated the flows radiometrically by the potassium-argon method. They found that magnetic reversals were synchronous around the globe. Geologists at Lamont Geology Observatory in Palisades, New York, studied the magnetization direction recorded in deep-sea sediment drill cores and correlated the dates with those of lava flows. The ambient ocean temperatures at different levels in the drill cores were inferred from fossil microorganisms present in the sediments (radiolarians, diatoms, foraminifera, and coccoliths). These tiny surface-dwelling animals and plants thrive at a variety of water temperatures; their proportions are sensitive indicators of climate. Ocean temperatures were also measured by using the oxygen-isotopic composition of the fossil foraminifera; those with a high proportion of the lightest oxygen isotope (oxygen 16) thrived at cooler temperatures.

By the mid-1970s geologists had distinguished four separate

temperature cycles with periods of 100,000 years, 42,000–43,000 years, 23,000–24,000 years, and 19,000–20,000 years. These cycles corresponded closely to variations in Earth's orbital eccentricity (a period of about 100,000 years), axial tilt (about 41,000 years), and precession (a major cycle with a 23,000-year period and a minor one with a 19,000-year period). With the publication of these data, the astronomical theory of climate variation achieved near universal acceptance.

The Antarctic ice sheet began forming about thirty million years ago, and massive ice sheets started to cover parts of the Northern Hemisphere about two and a half million years ago (fig. 5.4). During most of this latter period, ice sheets covered much of Canada but did not extend as far south as the Great Lakes. The ice waxed and waned with a period of 41,000 years, controlled by variations in the tilt of the rotational axis that ranged from 22.1° to 24.5°. By 700,000 years ago, the ice sheet advanced south of the Great Lakes and waxed and waned with a dominant period of 100,000 years, controlled mainly by variations in Earth's orbital eccentricity.

Geologists have found evidence for only a few other glacial epochs in the last billion years of Earth history—one in the Permo-Carboniferous period about 250–350 million years ago, one in the Siluro-Devonian period about 410 million years ago, one in the Ordovician period about 440 million years ago, and one in the late Precambrian about 700 million years ago. Thus, for most of the last billion years, Earth's climate has been relatively warm. It seems likely that orbital variations are not alone responsible for an ice age but that a trigger of some kind is also needed. One possibility is that higher albedo (overall reflectivity) due to increased cloud cover caused temperatures to decline until the astronomical variations could kick in. A more plausible trigger appears to be the latitudinal distribution of the continents. At the present time, Antarctica is situated over the South Pole, and the North Pole is partly surrounded by Greenland, northern Canada, and northern Asia. In the Permo-Carboniferous, most landmasses were joined together in a supercontinent dubbed Pangaea that

Figure 5.4. Earth from space. The arid regions of northern Africa and the Arabian Peninsula are cloudless. Antarctica is covered with a thick ice sheet. (Courtesy of NASA)

was centered over the equator. Nevertheless, parts of the continent (those areas containing portions of present-day South America, Antarctica, southern Africa, and India) extended to the South Pole. The circulation of ocean water near the poles was thereby impeded, and the water could not act to moderate temperature.

However, a different situation may have obtained in the late Precambrian sometime between 800 and 600 million years ago. The continents were relatively small, separated, and near the equator. Harvard University geologists Paul E. Hoffman and Daniel P. Schrag suggested that this situation created more rainfall that scrubbed carbon dioxide out of the air. The diminishment of this greenhouse gas lowered the global temperature sufficiently to allow the oceans to freeze over nearly completely, perhaps leaving only a narrow ice-free oceanic belt along the equator. This global glaciation event was termed "Snowball Earth" by geobiologist Joseph L. Kirschvink of the California Institute of Technology. Earth thawed only after about ten million years, when volcanic emissions (mainly from undersea volcanoes) had increased the concentration of carbon dioxide by three orders of magnitude over current levels. Carbon dioxide was able to achieve such high levels because of the near absence of rainfall in a world covered with ice. After the ice melted, global temperatures soared until rainfall was able to rid the atmosphere of most of the carbon dioxide.

Another variable that apparently affects global temperature is a minor variation in solar activity—changes in the number and size of sunspots, the occurrence of solar flares, and the extent of the corona. Although it is not clear how a slight decrease (on the order of 1 percent) in solar activity affects terrestrial climate, most researchers believe that there is a connection. From around 1450 to 1850, temperatures in the Northern Hemisphere were lower than normal. Glacial advances occurred in the Alps, Alaska, and the Sierra Nevada. Tide gauges in Germany and the Netherlands indicated correspondingly low sea levels. The River Thames in London and the Tagus River in Spain froze over during the winter months. This cold snap has been called the Little Ice Age. In the middle of this period, between 1645 and 1715, almost no sunspots were seen. On rare occasions when an astronomer observed one, it was cause for great excitement. During this interval there were few reports of auroras visible in northern Europe; when one was observed, many found the sight alarming. During total solar eclipses in this period, astronomers noted only a very small corona around the Sun.

Tree rings that formed at this time contain unusually high proportions of the heavy carbon isotope carbon 14. During periods of low solar activity, Earth's magnetic field expands; more cosmic rays bombard the atmosphere, producing secondary neutrons. These in turn are captured by atoms of nitrogen 14 in the atmosphere that then eject a proton and transform into radioactive carbon 14. This isotope combines with molecular oxygen to form carbon dioxide. During photosynthesis, trees incorporate the carbon 14–rich carbon dioxide and thus indirectly record low solar activity.

In 1986 John Martin of Moss Landing Marine Laboratories in California suggested that the cold temperatures at the time of the ice ages may have resulted in part from a complicated mechanism involving tiny marine plants called phytoplankton. In the iron-poor water near Antarctica, the Gulf of Alaska, and the equatorial Pacific, phytoplankton never bloom in large colonies as they do elsewhere in the ocean. Martin attributed this low fecundity to an iron deficiency. He hypothesized that airborne iron-bearing dust settled on the ocean surface during the ice ages and provided a missing nutrient to the plankton. The tiny plants proliferated and absorbed huge amounts of carbon dioxide from the atmosphere. The depletion of this greenhouse gas caused the climate to cool.

In 1993 and 1995, shortly after Martin's death from cancer, his associates conducted experiments on the open ocean. They dumped about 500 kilograms of iron (as acidic iron sulfate) in the equatorial Pacific. The phytoplankton bloomed and consumed hundreds of tons of carbon dioxide.

Nature may have conducted an analogous experiment 2.15 million years ago in the late Pliocene when a kilometer-size, metal-rich asteroid (the Eltanin meteorite) impacted the ocean in the southeastern Pacific near Antarctica. The projectile contained approximately 100 million metric tons of iron, some of which may have become available to the phytoplankton after the impact. The collision blasted more than a billion metric tons of water into the stratosphere, doubling the stratosphere's water content. In addition to the consumption of carbon dioxide by the plankton, the formation of high-altitude clouds may have reflected appreciable sunlight back into space and contributed to glacial cooling. In fact, the timing of this impact coincides with a major cooling event in the Northern Hemisphere after the onset of glaciation.

The ice ages demonstrate that the climate of Earth is linked to external influences—gravitational tugs from the Moon and Sun, solar activity, and the occasional impacts of asteroids. To paraphrase John Donne, who wrote in 1624, the middle of the Little Ice Age, "no [planet] is an island, entire of itself." Each is "a part of the main," an essential component of the solar system.

An understanding of Earth cannot be complete without detailed knowledge of how and when it acquired a moon. The origin of the Moon has been the subject of intense philosophical discussion. The modern consensus view of the Moon's origin involves the impact of a large body, perhaps several times the mass of Mars, with Earth more than four billion years ago. A more dramatic example of how Earth is connected to the cosmos cannot be found.

VI

ORIGIN OF THE MOON

O

Soon as the evening shades prevail,
The moon takes up the wondrous tale,
And nightly to the listening earth
Repeats the story of her birth;
While all the stars that round her burn,
And all the planets in their turn,
Confirm the tidings as they roll,
And spread the truth from pole to pole.
        —Joseph Addison, *Ode*

THE MOON has always held a special fascination for mankind. Its dominance of the nighttime sky, monthly cycle of phases, and mottled face have fueled speculation about its nature throughout history (figure 6.1). Before the return of the first lunar samples in 1969, Nobel laureate Harold Urey postulated that our satellite was composed of "genesis rocks" — matter essentially unaltered since the formation of the solar system. But the lunar rocks proved Urey wrong; the Moon, like Earth, is composed largely of igneous rocks formed from silicate melts. Its lowlands or maria (Latin for "seas") are covered with dark basalts that

Figure 6.1. The Moon as seen by the *Apollo 17* astronauts as they left lunar orbit and departed for Earth. (Courtesy of NASA)

flowed into place as molten lavas. The lunar highlands consist of lighter-colored rocks containing an abundance of calcium- and aluminum-rich silicates.

Lunar rocks generally range in age from three to four billion years; relatively few fragments have managed to survive from earlier times. Many lunar rocks had their isotopic "clocks" reset when they were pulverized and melted by the violent meteorite strikes that formed the Moon's innumerable craters. Unraveling lunar history from such complex, altered rocks is a difficult process.

A number of geophysical and geochemical constraints must be satisfied by successful theories of the Moon's origin. For example, Earth's mean density is 5.517 grams per cubic centimeter; if the compressional effects of gravity could be removed, Earth's density would be 4.03 grams per cubic centimeter, appreciably greater than the lunar value of 3.344. This difference immediately suggests that the Moon, relative to Earth, lacks metallic iron (whose density is about 7.9). However, the Moon's bulk density is very similar to the gravity-corrected value of 3.32 for Earth's outer mantle.

This similarity is apparent compositionally as well. The few patches of mantle rock exposed at Earth's surface, in concert with geophysical studies, show that the major minerals in our planet's upper mantle are the iron- and magnesium-bearing silicates olivine and pyroxene. During laboratory experiments, the partial melting of rocks rich in these two minerals produces basalts like those on Earth and the Moon.

A second constraint is imposed by the three naturally occurring oxygen isotopes, oxygen 16, 17, and 18 (all with eight protons but possessing eight, nine, and ten neutrons, respectively). Matter from different regions of the solar system contains these isotopes in varying proportions, providing a kind of planetary fingerprint that distinguishes one formative environment from another. Lunar samples contain oxygen isotopes in ratios indistinguishable from Earth's, implying that the two bodies formed in the same region of the solar nebula.

Third, despite the similarities noted above, the Moon is significantly depleted in two classes of elements relative to Earth and the common variety of meteorites known as chondrites. Elements of one class tend to melt or vaporize at relatively low temperatures. These are called volatiles and include hydrogen, chlorine, mercury, lead, and zinc. The Moon's apparent lack of water underscores this volatile depletion. (However, it is possible that some ice mixed with soil might reside on the floors of craters in permanent shadow at the lunar poles.) The other class of elements of low abundance in the Moon has an affinity for metallic iron. These elements are called siderophile (after the Greek words *sideros* meaning "iron" and *philos* meaning "loving") and include many metals such as nickel, cobalt, molybdenum, osmium, and iridium.

The Moon is odd in other ways, too. With the sole exception of the Pluto-Charon pair, our planet-moon system has the largest satellite with respect to its planet in the entire solar system. Moreover, the Moon's orbit is neither in the plane of the ecliptic nor in Earth's equatorial plane. Accounting for these facts is every bit as important as reconciling the geochemical evidence.

Before actually getting the lunar samples returned by the Apollo astronauts and the unmanned Luna spacecraft, planetary scientists had proposed three basic models of lunar origin:

1. gravitational capture by Earth of a satellite that formed elsewhere in the solar system;

2. fission from Earth, whereby the Moon was assembled from material that had previously been part of our young planet; and

3. binary accretion, which holds that the Moon and Earth formed side by side as a double planet.

The orbital planes of all the planets (except Pluto) are inclined to the ecliptic by no more than 7°, so the Moon's 5° inclination does not seem out of line and appears to be consistent with the capture hypothesis (figure 6.2). However, the capture of so large an object — not to mention the achievement of a nearly circular orbit in the process — is extremely improbable. Most bodies passing near Earth would simply be thrown into a new Sun-centered orbit. In order for a lunar-size object to have been snared by Earth, initially it must have occupied a heliocentric orbit very similar to ours. This coincidence seems very reasonable, considering the similar oxygen-isotope signatures. But it removes the most attractive feature of capture models — a simple explanation for the compositional differences between Earth and the Moon.

As an alternative, John Wood and Henri Mitler proposed a variation on the capture hypothesis in 1974. Their model relied on the fact that large objects in the solar system are compositionally differentiated. That is, at some point they melted enough so that the heavy metallic elements sank toward the center and collected into cores, leaving behind concentrations of lighter silicate minerals in the crusts and mantles of these bodies. (All the rocky inner planets seem to have undergone this basic first step in

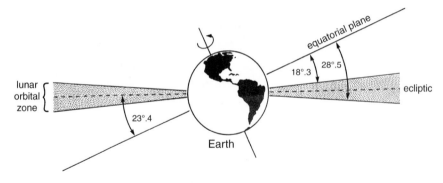

Figure 6.2. The plane of Earth's orbit around the Sun is called the ecliptic. Earth's equatorial plane is inclined 23.4° to the ecliptic. The Moon's orbit is inclined 5.1 the ecliptic and never makes an angle of less than 18.3° to Earth's equatorial plane.

planetary evolution.) Wood and Mitler suggested that a large number of smaller differentiated objects called planetesimals passed near Earth, where gravity-induced tides broke them into pieces. The silicate-rich mantles remained in captive orbits around Earth, while the more massive metallic cores escaped. The silicate debris then accreted into a single satellite.

However, this process seemed an inefficient way to build a large object. Wood later pointed out, "[w]ithout making special assumptions about the sizes and orbits of the [planetesimals], it becomes necessary to process an enormous mass of material . . . to obtain the makings of our Moon." Moreover, there would be a tendency for the metallic cores to return to the scene and join the accreting Moon, nullifying the chemical segregation achieved earlier.

In 1975 Mitler instead proposed that a single Mars-size object ventured too near Earth. It broke apart, and its silicate fraction was selectively retained in orbit. Although this model never gained wide acceptance, it had many of the features of the "Big Whack" model that achieved consensus about ten years later.

The first fission model of lunar origin was proposed in 1879 by George Darwin (the second son of Charles). He believed that our planet started out spinning very rapidly. When coupled with resonant effects induced by the Sun, this spin created enormous tidal bulges at Earth's equator. Given enough centrifugal force, a huge blob of matter ripped away to become the Moon. Darwin reasoned that tides raised by the new Moon would have slowed Earth's spin, transferring angular momentum to the Moon and moving it farther from Earth. Today, the Moon is receding at a rate of about 3 centimeters per year.

Ten years after Darwin's proposition, Osmond Fisher suggested that the Pacific Ocean basin was the scar marking the site where the Moon tore away. However, we now realize that Earth's crustal plates are in constant motion and the land-ocean configuration is ever changing. Even if Earth did spawn the Moon eons ago, our present landforms could not attest to it.

Darwin's fission model enjoyed wide acceptance for thirty years until Forest Moulton demonstrated that there was not enough angular momentum in the Earth-Moon system to create the necessary bulge. Besides, Harold Jeffreys later showed that such a bulge would be damped out by internal friction, making separation all the more unlikely.

In 1963, however, Donald Wise attempted to avoid these problems with a variation on the fission model. His primordial Earth rotated every 2.65 hours and, when its iron separated out and sank to form a core, the spin rate increased. Because its density was not uniform, Earth assumed the shape of a bowling pin, and the pin's "head" eventually broke free to form the Moon.

Another fission advocate, Alan Binder, argued as late as 1984 that known lunar rock types are geochemically consistent with a Moon extracted from Earth's mantle. He pointed out that because many binary stars with solar-type members may have formed by rotational fission, it is plausible that the Earth-Moon pair and the Pluto-Charon pair formed the same way.

Fission models offer several attractive characteristics that satisfy many of the geochemical constraints established during analysis of lunar samples. For example, a Moon ripped from Earth's mantle after differentiation had taken place would have the siderophile-poor composition now observed in lunar rocks. Similar

ratios among oxygen isotopes would be another obvious consequence, as would the matching densities of the Moon and Earth's
upper mantle. In addition, volatile gases, percolating outward to
form a primitive lunar atmosphere, could later have been sucked
away by Earth's stronger gravity, leaving the airless, bone-dry
world we see today.

An interesting variation on the fission hypothesis was offered
by the Australian theorist Ted Ringwood. In his scenario, the
early Earth differentiated and became so hot that metals and their
oxides boiled away, creating a thick, extended atmosphere. The
faster that Earth rotated, the more this atmosphere would have
collected along the equatorial plane. Eventually the gases cooled,
creating dust grains that ultimately collected into one large satellite. Most of the volatile elements escaped the Earth-Moon system altogether. Furthermore, if this occurred after Earth had already formed its iron core, then the dust grains that formed the
Moon would have already been depleted in siderophile elements.
Ringwood observed that "the relative abundances of siderophile
elements in Earth's upper mantle constitute a unique signature of
terrestrial origin."

However, Edward Stolper attacked Ringwood's "unique signature" argument because siderophile and volatile abundances similar to those in the Moon also occur in the basaltic meteorites
known as shergottites. These meteorites are now believed to have
come from Mars (but at the time were considered asteroidal); if
Mars (or an asteroid) could independently develop compositional
similarities to Earth, why couldn't the Moon have done so as
well?

A number of other objections have been raised against the fission hypotheses. All fission models require the Moon to have
formed in Earth's equatorial plane, yet the Moon's orbit is pres-
ently inclined between $18\frac{1}{2}°$ and $28\frac{1}{2}°$ to our equator. Fission proponents counter that a huge impact could have altered Earth's
rotational axis. Other planets and satellites exhibit a wide range
of spin orientations, they point out, so a lunar orbit inclined to
Earth's equator is not necessarily in conflict with fission models.

Figure 6.3. The first lunar meteorite to be identified, Allan Hills A81005. The rock consists of white anorthositic clasts (feldspar-rich rocks that are abundant in the lunar highlands) mixed with dark matrix material. The cube at left is 1 cm on an edge. (Courtesy of NASA)

A second objection concerns the siderophile elements. If the Moon formed from Earth's mantle after the terrestrial core formed, the siderophile abundances in lunar rocks should be similar to those in our mantle. However, the Apollo samples and lunar meteorites (figure 6.3) seem to be depleted significantly in these elements. It is likely that our satellite has a small core and that the Moon was partially molten at some point; siderophile elements would have migrated from the lunar crust and mantle to the core.

Finally, the problem of angular momentum has not disappeared. All fission models require the primordial Earth to have had much more angular momentum than any other terrestrial planet, and more than three times what the Earth-Moon system has now. Some angular momentum could have been lost over geologic time through the escape of atmospheric gases, but the dissipation of so much energy this way seems improbable.

Some lunar researchers advocated binary accretion models. The existence of a small lunar core and an orbit near the ecliptic are natural consequences of such models. The Russian scientist Elena L. Ruskol noted that silicate grains tend to fragment when they collide, whereas metallic grains tend to stick together. Therefore, collisions within the primordial solar nebula would produce fine silicate dust and heavier aggregates of metal; volatile elements would vaporize and escape altogether. Ruskol suggested that after growing to half its present mass, Earth gained a cloud of mostly silicate debris (metallic grains being more difficult to capture), which she termed the "circumterrestrial swarm." Ultimately, this swarm became the Moon—enriched in silicates and depleted in metals and volatiles.

Extending Ruskol's ideas, UCLA geochemists John Wasson and Paul Warren theorized in 1984 that many differentiated moonlets (similar to the planetesimals postulated by Wood and Mitler), each roughly 100 kilometers across, formed near Earth and collided with the swarm material. Their silicate mantles were chipped away and joined the cloud mass, while their metallic cores continued on into space.

These "double-planet" scenarios fail to explain why the other planets lack single, lunar-size satellites of their own. Although the Pluto-Charon system is even closer than the Earth-Moon system to being a double planet, other planet-moon systems are quite dissimilar. Jupiter has four large moons and at least two dozen smaller ones. Saturn has one large one and at least twenty-nine small ones. Uranus possesses only small satellites (twenty-one known as of summer 2001), while neighboring Neptune has one large one and more than half a dozen small ones. Mercury and Venus have no moons at all, although it is possible that Venus once had a moon but somehow lost it (perhaps by colliding with the planet billions of years ago). Mars has two moons, Phobos and Deimos, that are probably captured asteroids.

Considering all these variations, it seems likely that natural satellites resulted from a number of different processes driven by local circumstances. If this is correct, then we should not be averse to examine seemingly ad hoc models of lunar origin.

Perhaps the answer was staring us in the face all along. When we look at the face of the Man in the Moon, we are seeing enormous impact basins flooded with the volcanic rock basalt. These basins are huge craters formed after the Moon was struck billions of years ago by bodies tens of kilometers to about 100 kilometers in diameter. Even larger bodies must have struck Earth in the distant past.

William Hartmann and Donald Davis suggested in 1975 that a huge, Mars-size object (i.e., a body about eight times the mass of the Moon) collided with an already differentiated Earth. (Recent calculations indicate that the projectile is likely to have been two or three times more massive than Mars.) The impact could have blasted enough mantle debris into space to become reunited later as the Moon. However, dynamical calculations indicated that the debris would soon return to Earth's surface after making one geocentric orbit. In 1976 Al Cameron and William Ward proposed that if the material vaporized during the impact, it may have subsequently condensed in orbit around Earth.

After the pivotal Conference on the Origin of the Moon held in Kona, Hawaii, in 1984, these ideas gained wide acceptance as the "Big Whack" or "Giant Impact" model of lunar origin (figure 6.4).

As pointed out by astrophysicist Alan Boss in 1986, the formation of the Earth-Moon system is not a "special case" but is rather a natural consequence of planet formation. Because theoretical calculations indicate that the majority of the mass of accreting planetesimals resided in relatively few large objects instead of numerous small ones, giant impacts were inevitable.

In order to account properly for the angular momentum of the Earth-Moon system, the Mars-size projectile had to have struck Earth nearly tangentially at about 10 kilometers per second. Vaporized debris (admixed with solid and liquid) from the projectile and Earth's mantle were injected into orbit around Earth. A significant fraction of the volatile elements escaped, leaving the pro-

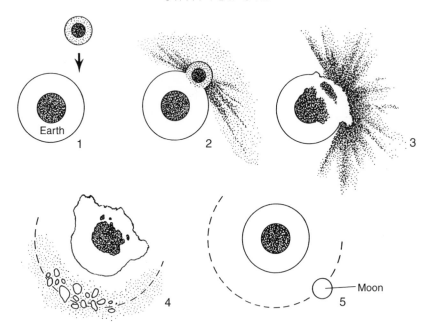

Figure 6.4. Cartoon of the Big Whack or Giant Impact model of lunar origin. A differentiated projectile a few times as massive as the planet Mars hits Earth tangentially. The metallic iron-nickel core of the projectile merges with that of Earth; silicate debris and vaporized material from the projectile and Earth's mantle go into orbit around Earth. This material eventually accumulates to form the Moon.

tolunar material depleted in these elements. Computer simulations of the impact indicate that the projectile's metallic core plunges into Earth, leaving the protolunar material in the circumterrestrial debris disk highly depleted in siderophile elements.

One problem with the Giant Impact model was that most simulations created a Moon in or within about 1° of Earth's equatorial plane. However, recent calculations by William Ward and Robin Canup of the Southwest Research Institute in Boulder, Colorado, demonstrate that gravitational resonance between Earth and the debris disk could increase its orbital inclination.

Although most lunar researchers have come to accept the Giant Impact model, many details remain to be worked out. Planning committees have urged NASA to return to the Moon (figure 6.5)

Figure 6.5. A $5 coin issued by the Republic of the Marshall Islands on 20 July 1989 commemorating the 20th anniversary of the first manned lunar landing.

and bring back material from a wide variety of sites. Because the Apollo and Luna rocks are all from the central nearside and the provenance of lunar meteorites is unknown, most terrain types remain completely unsampled. As expressed by lunar scientists Graham Ryder, Paul Spudis, and Jeff Taylor in 1989: "Literati still learn from Shakespeare; planetologists continue to learn from the Moon."

Part 3

SMALL BODIES, IMPACTS, AND RINGS

THE SOLAR system is more than a collection of eight or nine planets orbiting a central star. Trillions of other bodies are present. They range in size from submicrometer dust grains pushed around by solar radiation to cratered asteroids hundreds of kilometers in diameter. Planetary scientists discovered that Earth's crust preserves the remnants of impact craters and recognized that the craters on the Moon were formed by enormously energetic impacts. They came to realize just how important collisional processes were in the formation of planetary and subplanetary bodies. Groping for metaphors, scientists have likened the solar system to a pinball machine, a dartboard, a target, and a shooting gallery. The pinballs, darts, and bullets are errant asteroids and comets.

The four giant planets have numerous moons, some large, some small. Some moons were formed along with their planet in a manner analogous to the formation of the solar system itself. Other moons, especially distant small ones in inclined and eccentric orbits, are probably captured planetesimals. Each of the giant planets has a system of rings that harbors small moonlets.

In this section, we explore the nature of asteroids and meteorites, impact craters, and crater ejecta. We will examine the far-reaching geological and biological effects of a particular collision that occurred 65 million years ago between a 10-kilometer carbonaceous chondrite projectile and the Yucatán Peninsula in Mexico. We will look at recent cratering events in the solar system and see that impact cratering is an ongoing process. Finally, we will look back to the discovery of planetary rings and their accompanying moonlets and see how gravitational interactions make their existence possible.

VII

ASTEROIDS AND
METEORITES

O

Ah, Richard, with the eyes of heavy mind
I see thy glory like a shooting star
Fall to the base earth from the firmament.
Thy sun sets weeping in the lowly west,
Witnessing storms to come, woe and unrest;
Thy friends are fled to wait upon thy foes,
And crossly to thy good all fortune goes.

—William Shakespeare, *The Tragedy of King Richard the Second*

MAINLY residing in the region between Mars and Jupiter are planetesimals and their fragments that failed to accrete into a decent-size planet. These are the asteroids (typically tens to hundreds of kilometers in diameter) and their smaller cousins, the meteoroids. As they orbit the Sun, they often cross the orbits of other asteroids and occasionally smash into each other. Gravitational interactions with Jupiter send some of the bodies toward the inner solar system.

Some of the errant bodies plunge into Earth's atmosphere accompanied by sonic booms and whistling sounds. Those objects that survive to reach the surface become meteorites. A few have pelted living beings. About thirty stones fell in New Concord, Ohio, on 1 May 1860, and one of them killed a colt. Two stones fell in Sylacauga, Alabama, on 30 November 1954; one broke through the roof of a house, bounced off a radio stand, and hit a sleeping woman in the hip and hand. Hundreds of stones fell in Mbale, Uganda, on 14 August 1992; a 3.6-gram stone fell through the leaves of a banana tree and struck a boy on the head. The dead horse, bruised woman, and uninjured boy were connected to the asteroid belt in a very personal way.

Thirty-five years before the first asteroid was observed telescopically, the groundwork was laid by Professor Johann Titius of the University of Wittenberg, who developed a simple mathematical relation that closely approximated the known planets' distances from the Sun in astronomical units. A version of this relation follows: Begin with the numbers 0, 3, 6, 12, 24, 48, and 96 (each successive number after 3 being twice the preceding number). After adding 4 to each number and then dividing by 10, obtain 0.4, 0.7, 1.0, 1.6, 2.8, 5.2, and 10.0. Titius noticed in 1766 that if the number 2.8 was omitted from the series, the remaining entries were very close to 0.39, 0.72, 1.00, 1.52, 5.20 and 9.54 — the actual respective distances from the Sun, in astronomical units, of the planets Mercury, Venus, Earth, Mars, Jupiter, and Saturn. Titius did not attach much importance to these numbers and published them only as a footnote to a French astronomy book he was translating into German. Consequently, little attention was paid to his mathematical gimmick.

On 13 March 1781, five years after Titius's death, the great observational astronomer William Herschel discovered the planet Uranus in the constellation Gemini. Some months later the orbit of Uranus was calculated, and the planet's solar distance was found to be 19.19 AU. Johann Bode, director of the Berlin Observatory, pointed out that this value was quite close to the next number in Titius's series (19.6). Thus, it seemed possible that a heretofore undiscovered planet lay at 2.8 AU from the Sun, represented by the gap in the Titius sequence. From that point on this relation has commonly been called "Bode's Law."

Bode and Baron Franz von Zach of Seeberg began in 1796 to organize a group of twenty-four astronomers for the express purpose of finding the "missing planet" indicated by the relation to be between Mars and Jupiter. But it was not until September 1800 that even five astronomers could be assembled at Schröter's Observatory at Lilienthal to engage in the search. Von Zach called the astronomers the "celestial police" and assigned each member a particular area of the ecliptic to observe.

On the first night of the nineteenth century, 1 January 1801, in Palermo, Sicily, Guiseppe Piazzi was carefully recording stellar positions in Taurus while correcting a star catalogue. On the following evening, Piazzi noticed that one of the recorded stars had moved 4 arcminutes in retrograde, and on the next night, it had moved again by an equal distance. By 12 January, the object had changed its direction of motion and began moving prograde (eastward). Piazzi noted that its motion was far too slow for a comet but was altogether characteristic of a superior planet near opposition. The object also maintained a pinpoint starlike image. Meanwhile, a letter from von Zach was on its way to Piazzi. When Piazzi received the invitation to join the celestial police, he realized he had already discovered the "missing planet" and proposed a name for it, Ceres Ferdinandea. This was subsequently shortened to Ceres, the name of Sicily's patron goddess.

Piazzi sent a letter off to Bode, announcing the discovery of Ceres. Piazzi's observations were complete until 11 February, when illness prevented him from continuing, but his letter did not reach Bode until 20 March. By that time Ceres was far too close to the Sun to be picked up and was feared lost. The science of celestial mechanics was insufficiently sophisticated to enable the orbital elements of Ceres to be readily determined from Piazzi's scant six weeks of observations. However, Karl Gauss, a brilliant mathematician, gave himself the task of attempting to compute the orbit of Ceres from the limited data. He had been working on some new techniques for calculating orbits and welcomed the opportunity to test them. After determining the orbit and calculating new positions, Gauss communicated his results to Baron von Zach, who relocated Ceres on 7 December 1801.

On 28 March 1802 Wilhelm Olbers, a Bremen physician and amateur astronomer, discovered another body at approximately the same distance from the Sun as Ceres. He named it Pallas, after the goddess of wisdom. Karl Harding found Juno in 1804, and Olbers found Vesta three years later. (Juno and Vesta were respectively named after the queen of the gods and the Roman goddess of the hearth.) None of these four objects could be resolved telescopically as a disk. All appeared to be mere pinpoints of light.

Karl Hencke, a Berlin postmaster, discovered 5 Astraea in 1845 and 6 Hebe two years later. Since 1847, only the World War II year of 1945 has gone by without the discovery of additional asteroids, many being found accidentally by modern photographic methods. Clyde Tombaugh, the discoverer of Pluto, found the tracks of hundreds of asteroids in his photographic search for a trans-Plutonian planet. By summer 2001, well-defined orbits were known for about twenty-seven thousand asteroids, with this number doubling every two to three years. About ninety-nine thousand additional asteroids with less well defined orbits had also been catalogued by that time. There are probably a few hundred thousand main-belt asteroids large enough to be photographed from Earth.

Because more than one object had been found at about 2.8 AU, where Bode believed the missing planet resided, and because of the small sizes of these objects, Olbers suggested in 1803 that the asteroids were actually the remnants of an exploded planet. However, modern studies of the mineralogy and geochemistry of meteorites (the vast majority of which are from asteroids) have caused researchers to abandon the exploded planet theory.

In order to understand why this theory was abandoned, it is necessary to explore two related problems: the number of meteorite parent bodies and the identification of these bodies.

There are thirteen well-established chondrite groups with five or more members each. Each group has its own narrow range of mineral compositions, textural characteristics, bulk chemical composition, and bulk oxygen-isotopic composition. Mixtures among the chondrite groups are relatively rare, that is, there are relatively few examples of a fragment of a meteorite from one chondrite group residing within the matrix of a meteorite from another. Therefore, it is likely that each chondrite group was derived from a separate parent body. In addition to the established groups, there are about ten unique chondrites or chondrite grouplets (small groups with fewer than five members each) that were derived from separate parent bodies. There are also diverse groups of achondrites that formed on at least seven different parent bodies that experienced high temperatures and pervasive melting.

Iron meteorites constitute twelve main groups with distinct compositions. The narrow compositional ranges of each of the iron meteorite groups indicate that each was derived from a separate parent body. There are also numerous ungrouped irons and iron grouplets with distinct compositional characteristics, suggesting derivation from at least sixty different bodies. (The reason that there are so many iron meteorites is in part sociological; they look so different from terrestrial rocks that they are much more likely to be picked up in the field and brought to an expert for identification.)

Adding up the chondrites, achondrites, and irons, major groups, grouplets, and unique specimens, we find that the meteorites in our collections were derived from more than one hundred separate bodies. Any solid body in the solar system is a potential meteorite parent body. The list includes planets, moons, asteroids, and comets.

More than a dozen meteorites share the petrologic, mineralogical, geochemical, and isotopic characteristics of lunar rocks returned from the Moon by the Apollo astronauts and the unmanned Soviet Luna spacecraft. These rocks, known as lunar meteorites, were blasted off the surface of the Moon by energetic meteoroid impacts.

By summer 2001, there were eighteen volcanic meteorites thought to be Martian (figure 7.1). Some are about 1.3 billion years old, approximately the same age as that estimated for the

Classification of Meteorite Groups

Class Group	Characteristics
Chondrites	
Carbonaceous Chondrites	
CI	aqueously altered; chondrule-free; volatile-rich
CM	aqueously altered; small chondrules
CR	aqueously altered; metal-bearing
CO	small chondrules
CV	large chondrules; abundant CAIs
CK	large chondrules; darkened silicates
CH	microchondrules; metal-rich; volatile-poor
Ungrouped	(e.g., Coolidge)
Ordinary Chondrites	
H	high total iron
L	low total iron
LL	low total iron; low metallic iron
R Chondrites	highly oxidized; rich in ^{17}O
Enstatite Chondrites	
EH	high total iron; very reduced
EL	lower total iron; very reduced
Ungrouped	(LEW 87223)
Primitive Achondrites	
Acapulcoites	chondritic amounts of plagioclase and troilite
Lodranites	subchondritic amounts of plagioclase and troilite
Winonaites	IAB-silicate related
Ungrouped	(e.g., Divnoe)
Differentiated Meteorites	
Asteroidal Achondrites	
Eucrites	basalts
Diogenites	orthopyroxenites
Howardites	brecciated mixtures of eucrites and diogenites
Angrites	basalts rich in Ca-, Al-, and Ti-rich pyroxene
Aubrites	enstatite achondrites
Ureilites	olivine-, pyroxene-, and carbonaceous-matrix-bearing
Brachinites	olivine-, clinopyroxene-, and orthopyroxene-bearing
Martian Meteorites	
Shergottites	basalts and lherzolites

Classification of Meteorite Groups (*continued*)

Class Group	Characteristics
Nakhlites	Ca-pyroxene-bearing pyroxenites
Chassigny	dunite
ALH84001	orthopyroxenite
Lunar Meteorites	
Mare Basalts	flood basalts covering the maria
Impact Breccias	mixtures of lunar rocks plus some impact melt
Stony irons	
Pallasites	metal plus olivine; core-mantle boundary samples
Mesosiderites	metal plus basalt, gabbro, and orthopyroxenite
Ungrouped	(e.g., Enon)
Irons	
Magmatic Iron Groups	IC, IIAB, IIC, IID, IIF, IIIAB, IIIE, IIIF, IVA, IVB
Nonmagmatic Irons	Group IAB complex, IIE
Ungrouped	(e.g., Denver City)

Igneous rock types: pyroxenite (consists mainly of pyroxene); orthopyroxenite (consists mainly of orthopyroxene); gabbro (coarse-grained rock of basaltic composition); dunite (consists mainly of olivine); lherzolite (consists mainly of olivine, orthopyroxene, and clinopyroxene).

volcanoes on the Tharsis Ridge of Mars. Gas bubbles trapped within glass inclusions of one of these meteorites (EETA79001) have the same chemical and isotopic composition as the Martian atmosphere measured at the surface of Mars by the Viking lander in 1976. The presence of highly oxidized (i.e., ferric) iron in some mineral grains in the martian meteorites is consistent with the red dust that covers much of the planet; this dust contains the mineral hematite (Fe_2O_3), commonly known as rust.

Because the surface gravity of Mars is only 38 percent as strong as Earth's, hydrogen escapes from the top of the Martian atmosphere more readily than deuterium (heavy hydrogen atoms containing one proton and one neutron). The deuterium/hydrogen ratios in phosphate grains from these meteorites are about

Figure 7.1. A stamp (face value 12,500 Malagasy francs, FMG), equivalent to about U.S. $3.00) from the Republic of Madagascar showing an artist's rendition of the ALH84001 Martian meteorite. The souvenir sheet also shows an

three to five times higher than in terrestrial rocks, consistent with the ratio measured in the Martian atmosphere.

All the rest of the meteorites, encompassing tens of thousands of individual specimens, are believed to come from asteroids. This inference is based on nine major links between meteorites and asteroids:

1. High-energy nuclear particles known as cosmic rays penetrate silicate rock to depths of about 1 meter, smashing into atoms in the rock and transforming some of them into radioactive isotopes such as neon 21 and argon 36. The decay of these isotopes can be used as a clock, timing the period the meteoroids existed in space as meter-size objects. The cosmic-ray-exposure ages of stony meteorites range from thirty thousand to seventy million years. This is far too short a time for meteoroids to have traveled to Earth from the vicinity of another star; such interstellar trips would probably take at least several hundred million years. This indicates that meteorites are products of our solar system. Specifically, these cosmic-ray-exposure ages match the typical times expected from theoretical calculations for objects to travel to Earth from the asteroid belt.

2. The solar corona (the outer atmosphere of the Sun visible during total eclipses) expands into interplanetary space as the solar wind, carrying with it noble gases such as hydrogen, helium, neon, and xenon. Bodies near the Sun are blasted with high concentrations of these noble gases; more distant bodies acquire lower concentrations of solar gas. The abundance of solar-wind-implanted noble gases in some meteorites is consistent with their acquisition at about 3 AU from the Sun—that is, in the center of the asteroid belt.

3. The very presence of solar-wind-implanted noble gases and solar-flare particle tracks in some meteorites indicates that they were exposed at the surface of an airless body (figure 7.2). This is because collisions between solar-wind particles and the molecules in a planet's atmosphere shield the planet surface from exposure to the solar wind. Thus, meteorites containing solar-wind gases cannot have been derived from objects with substantial atmos-

Figure 7.2. The Cangas de Onis H-chondrite regolith breccia meteorite. This rock consists of light-colored clasts of metamorphosed H chondrites embedded in a darker-colored matrix. The matrix contains high concentrations of noble gases such as helium and neon implanted from the solar wind. Scale at bottom is in centimeters. (Courtesy of Jeff Taylor, University of Hawaii)

pheres such as Venus and Saturn's moon Titan. Asteroids are too small and have too little gravity to have retained an atmosphere; they are likely candidates for the parent bodies of meteorites containing abundant solar gas.

4. Many meteorites are made up of broken fragments. Some of the fragments are remnants of projectiles that impacted the parent asteroid at fairly low relative velocities (figure 7.3). The high proportion of broken fragments in these meteorites is consistent with the dense cratering observed on asteroids that have been photographed by passing spacecraft (i.e., Ida, Gaspra, Mathilde, and Eros).

5. Repeated gravitational perturbations by Jupiter cause asteroids in certain orbits to change their orbital characteristics and assume Earth-crossing orbits. There are presently estimated to be

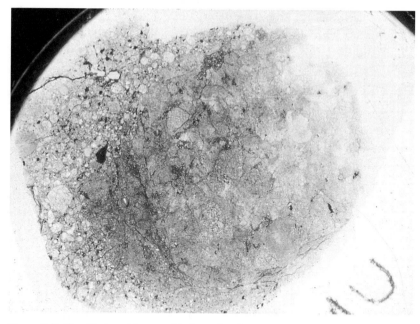

Figure 7.3. The H-chondrite regolith breccia Dimmitt, shown here in thin section, contains some foreign fragments of other meteorite varieties. The small chondrules of the Dimmitt host are at left; a shocked LL5 chondrite fragment with large chondrules occupies the right two-thirds of the section. Also present in the Dimmitt host is a small black carbonaceous chondrite fragment. The entire section is 16 mm across.

about nine hundred asteroids greater than 1 kilometer in diameter in Earth-crossing orbits. Calculations indicate that 7 percent of them will eventually strike Earth.

6. Analysis of the size and composition of meteorite metal grains and of plutonium fission tracks in phosphate grains indicates that many meteorites cooled from high temperatures at rates between 1 and 100°C per million years. The slowest rates correspond to those expected from material near the centers of rocky bodies 100–300 kilometers in diameter. This size interval matches that of many large asteroids.

7. If small objects and large objects are heated to the same temperature, the small objects cool more rapidly because they have higher surface-area/volume ratios. In other words, relative

to their trifling volumes, there is a lot more surface on small bodies from which heat can radiate away. This is why large rocky bodies like Earth still have enormous reserves of internal heat four and a half billion years after they formed, while small bodies like asteroids cooled completely shortly after formation. The four-and-a-half-billion-year age of most meteorites thus indicates that they were derived from small, asteroid-size bodies. This conclusion is consistent with the results of a survey of asteroids larger than 10 kilometers in diameter; the survey found that 95 percent of these objects have diameters less than 200 kilometers.

8. Different minerals absorb and reflect light in characteristic ways at different wavelengths. Laboratory studies of the spectral reflectivities of certain meteorites closely match those made telescopically of certain asteroids. These studies reveal that many meteorites have compositions similar to those of some asteroids (figure 7.4).

9. Since the late 1950s, six chondritic meteorite falls (Príbram, Czech Republic, 7 April 1959; Lost City, Oklahoma, 3 January 1970; Innisfree, Alberta, 5 February 1977; Peekskill, New York, 9 October 1992; Tagish Lake, British Columbia, 18 January 2000; Morávka, Czech Republic, 6 May 2000) were photographed or videotaped as they plunged through the atmosphere toward the ground. Reconstruction of their orbits reveals close similarities to those of typical Earth-crossing asteroids (figure 7.5).

These observations demonstrate that the vast majority of meteorites are from asteroids.

There is a correlation between the compositions of the asteroids and their distances from the Sun. A clustering of the bright, highly reflective, stony asteroids occurs in the inner portion of the belt (at 2.3 AU), and with increasing solar distance a higher percentage of dark objects can be found. Beyond 3 AU, more than 95 percent of the asteroids seem to be rich in organic materials; many are probably rich in water as well. These compositional differences must have arisen from the thermal gradient in the solar nebula over four and a half billion years ago. It seems natural that the inner (terrestrial) planets and innermost asteroids would

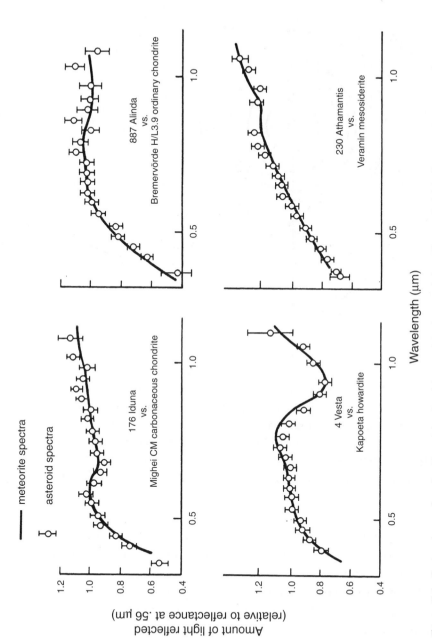

Figure 7.4. Comparison of meteorite and asteroid spectral characteristics. The asteroidal data points determined telescopically closely match the meteorite curves determined in the laboratory. These correspondences indicate that the surfaces of the asteroids are similar in composition to the corresponding meteorite varieties.

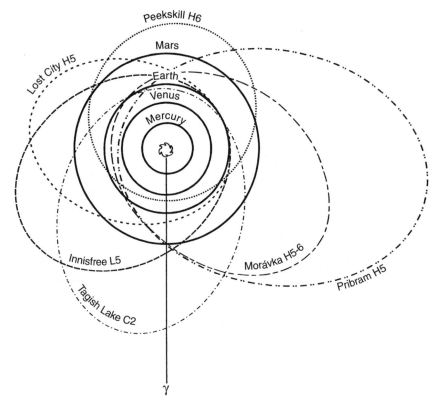

Figure 7.5. Six chondrites in our meteorite collections were photographed or video-taped as they entered Earth's atmosphere. The orbits of these meteorites are similar to those of typical Earth-crossing asteroids. The line with the γ symbol is a directional marker indicating the vernal equinox, one of two points in space where the ecliptic intersects the celestial equator. In the case of the vernal equinox, it is the point that the Sun passes through around 21 March at the beginning of spring.

be rocky while at the outer edge of the asteroid belt temperatures must have been low enough to prevent the organic compounds from volatilizing.

The main-belt asteroids must still be approximately at the same distances from the Sun at which they originally accreted, although some mixing has subsequently occurred from collisions and gravitational perturbations. These data alone render the shattered planet hypothesis untenable, because such a cataclysmic explosion would have distributed the planet's fragments isotropically throughout the solar system. There could not now be certain regions where the minor planets were predominantly stony and others where they were carbonaceous. Furthermore, meteorites that were melted by igneous processes cannot be from the same parent bodies as unmelted, primitive chondrites. The oxygen-isotopic compositions of different meteorites are incompatible with derivation from a single body.

In 1918 the Japanese astronomer Kiyotsugu Hirayama pointed out that asteroids could be grouped together by the similarities of their orbital elements. The computed eccentricities, semimajor axes (equivalent to their mean heliocentric distances), and orbital inclinations all appeared to be closely distributed around certain values. His groups have come to be known as "Hirayama families." Each family can be thought of as the debris remaining after the break-up of a parent minor planet (figure 7.6). After a major collision that disrupted the parent body, the fragments dispersed and distributed themselves along the original body's orbit. Thus, instead of all the asteroids being fragments of an exploded planet, the model fits the origin of individual asteroid families.

In 1867 astronomer Daniel Kirkwood noticed that a plot of the number of asteroids versus their semimajor axes was not smooth; instead there were gaps containing few or no asteroids (figure 7.7). These gaps have come to be known as Kirkwood gaps; they occur at 2.1 AU, 2.5 AU, 2.8 AU, and 3.3 AU.

Kepler's third law states that P^2 is proportional to a^3, where P^2 is the square of the period in years of an object orbiting the Sun and a^3 is the third power of the semimajor axis of an orbiting object. Kirkwood found that these gaps correspond to orbital periods of 3.0, 4.0, 4.7, and 6.0 years, respectively. Because Jupiter's orbital period is about 12 years, the gaps correspond to simple fractions of Jupiter's orbital period: $\frac{1}{4}$, $\frac{1}{3}$, $\frac{2}{5}$ and $\frac{1}{2}$, respectively. An asteroid with a period of $\frac{1}{4}$ of Jupiter's period is close to Jupiter every four times it orbits the Sun; furthermore, this proximity occurs at the same point in the asteroid's orbit. Repeated gravita-

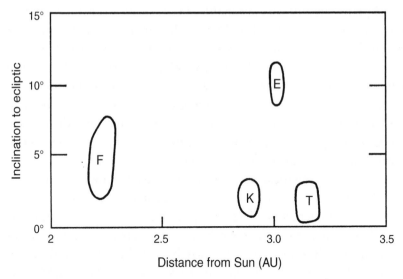

Figure 7.6. When the inclinations of asteroids relative to the ecliptic are plotted against the asteroids' mean distances from the Sun, several clumps of data points appear on the diagram. These clumps represent asteroids of similar orbital characteristics and are called Hirayama families. Although there may be a few unrelated asteroids within each clump, each true family member was derived from the same disrupted asteroid. The four major families shown here are the Flora (F), Koronis (K), Eos (E), and Themis (T) families, each named after the largest (and brightest) member.

tional tugs from Jupiter nudge the asteroid out of its orbit. In some cases, asteroids cleared from the Kirkwood gaps develop eccentric orbits and are more likely to collide with planets in the inner solar system. Chips from these asteroids can move into Earth-crossing orbits and fall to Earth as meteorites.

The August 1993 flyby of the 56-kilometer-long, heavily cratered asteroid 243 Ida (figure 7.8) by the *Galileo* spacecraft revealed the presence of a small, egg-shaped moon measuring 1.2 × 1.4 × 1.6 kilometers in size. This was the first moon discovered orbiting an asteroid. The moon was named Dactyl after the Dactyli, Greek mythological beings that inhabited Mount Ida,

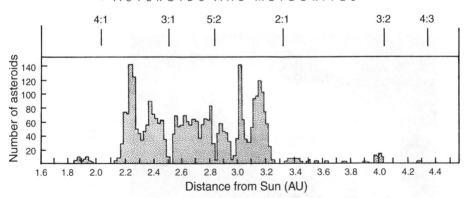

Figure 7.7. Plot of the number of individual asteroids at different distances from the Sun. Gaps in the diagram (e.g., at 2.5 and 3.3 AU) match resonances with Jupiter; asteroids at these distances were perturbed into different orbits by Jupiter's enormous gravity. However, at certain resonances (e.g., 3:2 and 4:3), asteroids are concentrated by Jupiter's gravitational influence. The resonance numbers indicate the number of orbits an asteroid completes relative to the number of orbits completed by Jupiter in the same period of time. For example, an asteroid at the 3:1 resonance would orbit the Sun exactly three times for every single time that Jupiter orbits the Sun.

where the infant Zeus was raised. In November 1998 astronomer William J. Merline and colleagues used the 3.6-meter Canada-France-Hawaii Telescope on Mauna Kea to image a small moon orbiting the 200-kilometer-wide asteroid 45 Eugenia. The moon orbits Eugenia every 4.7 days. By summer 2001, about ten asteroid-moon pairs had been discovered. The origin of asteroidal moons is a mystery. Although they may have been captured by their parent asteroid during a collision that formed a Hirayama family, theoretical calculations suggest that they would have been destroyed or stripped away within a few hundred million years. Alternatively, they could have accreted in orbit around their parent asteroid from material blasted off the parent during a major cratering event.

Details of the history of asteroids continue to be revealed by laboratory studies of meteorites and spacecraft studies of asteroids. But the greatest insights probably await sample-return missions from selected asteroids. At that point, meteorite researchers

Figure 7.8. Composite of photographs of three asteroids encountered by the *NEAR* and *Galileo* spacecraft: *left*, 253 Mathilde (50 × 50 × 70 km); *center*, 951 Gaspra (11 × 12 × 19 km); *right*, 243 Ida (21 × 24 × 56 km). (Courtesy of NASA)

will become field-workers, and the science of meteoritics will mature into asteroid geology. As material from known sources becomes available, it may be possible to answer one of the most puzzling questions of planetary science: What heated the asteroids? Possible answers are explored in the next chapter.

VIII

WHAT HEATED
THE ASTEROIDS?

O

Crushed and cratered, shattered, pelted,
Shocked and fractured, sintered, melted,
Blackened, jumbled, porous, blocky,
Rent, fragmented, vuggy, rocky,
Slowly spinning in the void;
It's a battered asteroid!
　　—Alan Rubin, *Waltzing 253 Mathilde*

METEORITES of bewildering variety fall to the ground. A few are iron, most are made mainly of silicate minerals, and a few are mixtures of both. About 85 percent of the meteorites observed to fall are chondrites, primitive objects similar in composition to the nonvolatile portion of the Sun's surface. These meteorites derive their name from the submillimeter, once-molten silicate spherules (chondrules) that most of them contain. In addition to chondrules, chondrites contain other silicate grains as well as metal, sulfide, oxide, and phosphate.

Approximately 10 percent of the chondrites (the so-called type-3 chondrites) have never experienced temperatures higher than 400–600°C. They are nearly pristine agglomerations of discrete inclusions derived from the solar nebula — the cloud of gas and dust that gave rise to the solar system. Type-3 chondrites are mineralogically unequilibrated: adjacent grains of the same mineral phase may have quite different compositions. Such meteorites are also texturally unrecrystallized; adjacent grains have not grown into one another. Type-3 chondrites contain moderately high abundances of noble gases and water. Their matrices contain tiny presolar grains that formed in the outer atmospheres of other stars and fell into the early solar system from interstellar space.

However, most chondrites do not possess these characteristics. The types -4 to -6 chondrites have experienced prolonged periods at elevated temperatures (600–950°C). These thermally metamorphosed rocks have equilibrated mineral compositions: adjacent grains of the same mineral phase have essentially identical compositions. The textures of these chondrites show significant recrystallization: minerals have overgrown and replaced one another, and grain sizes have become more uniform (figure 8.1). Volatile elements and compounds have been partly baked out; tiny presolar grains trapped within the matrix have been destroyed.

Meteorites that experienced temperatures even higher than those affecting the type-6 chondrites were melted completely. During melting, chondrites form two immiscible liquids: a low-density silicate melt and a high-density metal-sulfide melt. The silicate melts crystallize into igneous rocks known as achondrites (i.e., "no chondrules"). Many are fine-grained rocks called basalts (figure 8.2), similar to basalts from Earth's ocean basins and the volcanically flooded lunar impact basins (i.e., the maria). Iron meteorites formed by crystallization of the metal-rich melts. Stony-irons, which are composed of half iron and half silicate, comprise two main groups: pallasites, which formed at the boundary between the metal core and silicate mantle of a melted and differentiated body (figure 8.3); and mesosiderites, which, as explained in chapter 9, may have formed near the basaltic surface of one differentiated body after mixing at low velocities with portions of a metal core from another differentiated body.

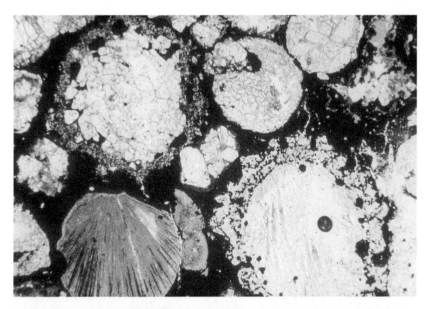

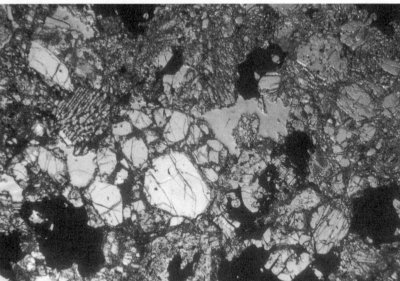

Figure 8.1. Comparison of the textures of a primitive and a metamorphosed ordinary chondrite. The top photomicrograph shows chondrules in a thin section of the unequilibrated LL3.0 chondrite Semarkona. The bottom photomicrograph shows the recrystallized texture in a thin section of the metamorphosed H5 chondrite Allegan. Chondrule outlines in Allegan are difficult to discern. The field of view in both images is 2.9 mm.

Figure 8.2. A portion of a basaltic clast in a thin section of the Pasamonte eucrite meteorite. The elongated needlelike minerals are plagioclase feldspar; the large, more rectangular grains are calcium-rich pyroxene. Field of view is 2.5 mm. (Courtesy of Jeff Taylor, University of Hawaii)

About 92 percent of meteorite falls have either been melted like the basalts, irons, and stony-irons or appreciably metamorphosed like the types -4 to -6 chondrites. Therefore, the bodies these meteorites came from must have been significantly heated. Since the vast majority of meteorites come from asteroids, many of the asteroids able to send materials to Earth must have been hot at one time. This same conclusion was reached by some asteroid researchers who attacked the problem from the other direction. Their spectral reflectivity studies suggest that many of the asteroids closer to the Sun than about 3 AU were metamorphosed or melted. The question then arises: What could have heated so many different asteroids?

Some potential heating mechanisms can be eliminated immediately. Nuclear fusion — the transmutation of hydrogen nuclei into helium nuclei — is the process that powers the stars. Tem-

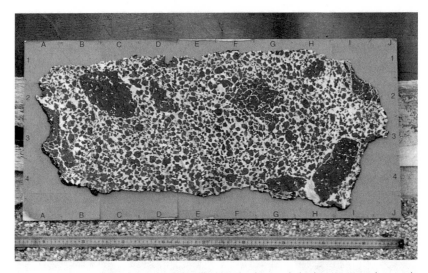

Figure 8.3. Large slab of the Esquel pallasite. The large, dark clasts are made mostly of olivine. They are fragments of the lower mantle of a differentiated asteroid that mixed with metal from the top of the core. The slab is 90 cm across; a meter stick is shown for scale. The slab was on loan to UCLA, courtesy of Bob Haag. (Photograph courtesy of John Wasson and Finn Ulff-Møller, UCLA)

peratures in excess of fifteen million degrees are required for this process to operate. Such temperatures are generated in the centers of stars by the gravitational pressure caused by more than a trillion trillion metric tons of gas. In contrast, Jupiter, the most massive planet in the solar system, would need to be about forty to sixty times more massive to achieve the enormous core pressures required for nuclear fusion. Ceres, the largest asteroid, is 1.6 million times less massive than Jupiter. This is far too small to allow nuclear fusion.

It is also impossible for the asteroids to have been heated by long-lived radionuclides such as potassium 40, thorium 232, uranium 235, and uranium 238. These isotopes have half-lives ranging from seven hundred million years for uranium 235 to fourteen billion years for thorium 232. Although Earth's mantle is heated by the decay of these isotopes, chondritic bodies smaller than about 1,500 kilometers across (a size range that includes all of the asteroids) that contain these isotopes in the parts per million to parts per billion range (as expected for chondrites) could not be. They would lose heat too rapidly to reach high temperatures because of their high surface-area/volume ratios. In other words, in asteroid-size bodies, heat is lost faster than it is generated by the decay of long-lived radionuclides.

Furthermore, most meteorites were metamorphosed or melted four and a half billion years ago. This old age is inconsistent with heating by long-lived radioactive isotopes because bodies heated in this way would not reach their maximum temperatures until hundreds of millions of years after they had formed. In other words, if meteorites were heated in this manner, their formation ages would be far younger than four and a half billion years. (Although some meteorites do have younger ages, these are usually attributed to late-stage shock resetting of their isotopic clocks.)

Another implausible asteroidal heating mechanism is tidal heating. This is frictional heating caused by uneven gravitational tugs on a body; significant tidal heating occurs only in satellites such as Io around large planets such as Jupiter.

Exothermic chemical reactions, such as occur when sodium hydroxide is added to water, are also unlikely to heat asteroids. Astrophysicist Don Clayton suggested that chemical energy in interstellar dust grains might not have been released until small rocky bodies had accreted. He proposed that exothermic chemical reactions could have thermally metamorphosed chondrites and melted achondrites and irons. However, this model fails to explain the existence of the unmetamorphosed, unrecrystallized type-3 chondrites. These rocks formed from the same material and at the same time as the metamorphosed chondrites but somehow managed to remain cold. This is an apparently fatal flaw.

There are three principal mechanisms that have been widely discussed as being potentially capable of heating asteroids: (1) decay of short-lived radionuclides, (2) electromagnetic induction in the protosolar wind, and (3) collisional heating.

Most meteorite researchers believe that the decay of short-lived radionuclides is probably responsible for heating the asteroids. There are two candidate isotopes: aluminum 26, with a half-life of 720,000 years, and iron 60, with a half-life of 1.5 million years. Because recent research has indicated that there was probably too little iron 60 available at the time of accretion to have caused much heating, most researchers have focused on aluminum 26. This radioactive isotope decays into a stable isotope — magnesium 26. Because some magnesium 26 was already present in the rock, the former presence of aluminum 26 is inferred from excess amounts of magnesium 26. Such excesses have been found in many refractory inclusions (small objects composed of minerals that melt at high temperatures) in chondrites. Because some refractory inclusions lack magnesium 26 excesses, they may have formed before aluminum 26 entered the solar system or several million years after the aluminum 26 had decayed away. Alternatively, if aluminum 26 was not uniformly distributed in the solar system, inclusions lacking magnesium 26 excesses may simply have formed in aluminum 26–free regions. This uncertainty has important consequences for the viability of aluminum 26 as a heating mechanism for asteroids: if aluminum 26 was uniformly distributed, it would have been available to heat asteroid-size bodies at different heliocentric distances; if it was heterogeneously distributed, it may have been available to heat bodies in certain regions but not in others. The question of the uniformity of the distribution of aluminum 26 in the early solar system has not been resolved.

Small inferred initial amounts of aluminum 26 in basaltic meteorites were recently reported. This suggests that the decay of aluminum 26 may have played a major role in melting chondritic asteroids and forming achondrites.

Another widely discussed asteroidal heating mechanism is electromagnetic induction. The early Sun may have gone through a highly variable T-Tauri phase in which an intense ionized solar wind with an embedded magnetic field streamed out of the Sun at velocities of several hundred kilometers per second. As this wind swept through the solar system, it encountered asteroids that had a small amount of inherent electrical conductivity. Motion of the magnetic field lines induced a current in the asteroids, causing some heating to occur.

There are reasons to question the effectiveness of this heating mechanism. An intense solar wind, ten million to one hundred million times more intense than the present wind, is required for electromagnetic induction to work, and it is far from certain that the Sun ever possessed a wind of such intensity. Furthermore, T-Tauri stars have been observed to blow off much of their mass at their poles, far away from any asteroids that may be residing in the equatorial accretion disk. Because of such uncertainties, most researchers have abandoned electromagnetic induction.

Another plausible mechanism for heating the asteroids is by meteoroid collision. Spacecraft images of asteroids 951 Gaspra, 243 Ida, 253 Mathilde, and 433 Eros (figure 8.4) show that these bodies are heavily cratered, indicating that collisions among asteroids were frequent. This observation is consistent with the evidence that many meteorites have been shocked. Shocked meteorites contain damaged mineral grains, broken clasts, patches of melted silicate, and veins of silicate glass enclosing small melted spherules of metal and sulfide. In addition, some chondrites contain minerals that formed under transient high pressures typical of shock. A few shocked chondrites even contain fragments of projectiles of different chondrite groups embedded within them. Many chondrites contain fine-grained silicate clasts that formed by impact melting early in solar-system history and were later metamorphosed along with the rest of the meteorite. The presence of these clasts implies that the impact events that formed them preceded or accompanied metamorphic heating.

Collisional heating of chondrite parent bodies is also supported by a correlation among the different chondrite groups between metamorphism and shock: those groups that have few or no metamorphosed members (such as the carbonaceous chondrites) also have few or no shocked members, whereas those groups that have many metamorphosed members (e.g., the ordinary chondrites and enstatite chondrites) also have many shocked members.

Figure 8.4. A photomosaic of six digital images of the north polar region of asteroid 433 Eros made by the *NEAR-Shoemaker* spacecraft from an orbital altitude of 200 km. The crater in shadow at the top is 5.3 km in diameter. (Courtesy of NASA and The Johns Hopkins University Applied Physics Laboratory)

Despite these observations, many meteorite researchers have discounted collisional heating models because calculations indicate that most asteroidal bodies are too small to have accreted material with enough kinetic energy to have caused widespread heating of the target. Computer modeling suggests that over the lifetime of an asteroid (the period between its formation and the time it is disrupted by collisions), there will be only small amounts of melt produced in comparison to the amount of impact-generated fragmental debris. Theoretical calculations and extrapolations from laboratory shock experiments imply that the low escape velocities of asteroids would permit much of the most strongly heated debris to escape during impact-crater formation. Finally, radiative heat loss between impact events would hamper cumulative temperature increases from successive impacts.

Although these arguments are sound, they are applicable mainly to cohesive asteroids of low porosity. The heating efficiency of impacts increases for more porous bodies. Recent data suggest that many asteroids are giant piles of rubble. Radar observations of near Earth asteroids show that many appear to be low-strength aggregates of boulders. Theoretical studies of the collisional evolution of asteroids indicate that for a large range of impact energies, collisions can disrupt asteroids but fail to disperse most of the fragments. Poorly consolidated rubble piles of high porosity and low density would result. This conclusion is consistent with the low densities observed for the asteroids 45 Eugenia (1.2 grams per cubic centimeter), 253 Mathilde (1.3 grams per cubic centimeter) and 243 Ida (2.5 grams per cubic centimeter). (For comparison, many silicate minerals have densities of 3.0–3.2 grams per cubic centimeter, and water has a density of 1.0 gram per cubic centimeter). The inferred porosities of these asteroids are also high: e.g., 50 percent for Mathilde and 30 percent for Ida; that is, half the volume of Mathilde and nearly a third of the volume of Ida seem to be empty space. Orbital images of 433 Eros reveal linear features such as squared-off crater walls, grooves, and ridges, suggesting that this asteroid might not be a rubble pile. However, its low bulk density (2.67 grams per cubic centimeter) implies significant internal porosity (10–30 percent).

When the NEAR (Near Earth Asteroid Rendezvous)* spacecraft flew by Mathilde in June 1997, it photographed about 50 percent of the asteroid's surface. These images show that there are five large craters, 19–33 kilometers in diameter, that are comparable in size to the mean radius of Mathilde (26.5 kilometers).

---

*On 14 March 2000 NASA renamed the spacecraft NEAR-Shoemaker as a tribute to planetary scientist Gene Shoemaker, who died in a car accident in 1997 while investigating impact craters in the Australian outback.

All five craters lack significant ejecta blankets, suggesting that this is a characteristic feature of porous asteroids. The absence of ejecta blankets on Mathilde is consistent with laboratory experiments that indicate only small amounts of material are ejected during cratering events on porous bodies. Most of the impact energy goes into crushing and heating the target material instead of ejecting debris. Therefore, if many asteroids are heavily cratered rubble piles (and since rubble piles are efficient at retaining heat), collisions may be a viable mechanism for heating asteroids.

It must be kept in mind, however, that the different heating mechanisms are not mutually exclusive. There may be different mechanisms operating more efficiently at different times or at different distances from the Sun. (This is obviously true for short-lived radionuclides, which decay away after a few million years.) There are also several important unanswered questions about heat sources. Because there are probably some asteroids that are not rubble piles, they are unlikely to have been heated significantly by collisions.

If both aluminum 26 and collisions heated asteroids, which was the dominant mechanism? Did this dominance reverse at different heliocentric distances? Answers to these questions await advances on a number of fronts: additional remote sensing of asteroids, more-sophisticated laboratory experiments of impact cratering, increasingly detailed computer modeling of heat sources, continued laboratory investigations of meteoritic components, and sample-return missions from asteroids.

An unusual meteorite group that shows evidence for both extensive heating and shock metamorphism is the set of mesosiderites. Detailed studies of these enigmatic meteorites could help resolve the nature of asteroidal heating. Mesosiderites are examined in the next chapter.

IX

MESOSIDERITES:
BIOGRAPHY OF A SHOCKED
AND MELTED ASTEROID

O

> Ay me! What perils do environ
> The man that meddles with cold iron;
> What plaguy mischiefs and mishaps
> Do dog him still with after-claps!
> For though Dame Fortune seem to smile
> And leer upon him for a while,
> She'll after shew him in the nick
> For though Dame Fortune seem to smile
> Of all his glories, a dog trick.
> —Samuel Butler, *Hudibras*

MESOSIDERITES comprise a strange group of a few dozen meteorites that have caused considerable consternation among the scientists who study them. The puzzle of the meso-siderites resides in their stony-iron composition; their name derives from the Greek *mesos*, meaning "middle" (loosely, "half"), and *sideros*, meaning "iron." In terms of mass, mesosiderites con-

sist of about 50 percent metallic iron-nickel and 50 percent silicates (figure 9.1). This is a curious mixture for a meteorite that
has undergone melting: iron and nickel are dense metals that are
usually found in the core of planetary bodies, whereas the comparatively buoyant silicates in a mesosiderite are usually found on
the surface and in the mantle of such bodies. The oddness of
finding this metal and silicate concoction in a meteorite that was
once molten is analogous to discovering a blend of cashews and
steel ball bearings on the side of the road. It is a combination that
must certainly have an interesting story behind it.

Some investigators have abandoned the thought of explaining
the existence of these meteorites, occasionally referring to them
as "messy-siderites." But because celestial bodies such as asteroids have remained geologically inactive for billions of years, the
composition and structure of mesosiderites hold vital information

Figure 9.1. Slab of the Emery mesosiderite showing a large, dark pyroxene-rich clast
at one side and a round white metal nodule at the other. The slab is 10 × 16 cm in
size. (Courtesy of Marty Prinz, American Museum of Natural History)

about some of the complex processes that planetary bodies must have experienced early in solar-system history.

Over the past three decades, investigators have analyzed thousands of meteorites, including the mesosiderites, in an attempt to reconstruct this history. However, it was not until detailed geochemical and mineralogical studies were made that the evolutionary history of the mesosiderite parent asteroid could be reconstructed. NASA scientist David Mittlefehldt and I have compiled information about these meteorites that has allowed us to outline the evolutionary history of their parent asteroid. This is in fact the biography of that asteroid. It is a body with a complicated geological history, a body that has been shocked and melted, a body that has suffered repeated impacts from other asteroids.

Using geochemical and isotopic analyses, we have discerned five major episodes in the evolutionary history of the mesosiderite parent asteroid.

Accretion was the first major event in the history of all of the planets and asteroids, including the mesosiderite body. From isotopic analyses of refractory inclusions in chondrites, we infer that accretion occurred about 4.56 billion years ago as dust grains clumped together to form planetesimals and these in turn collided at low velocities to form asteroid-size bodies. Astrophysical models suggest that only a relatively short time span, probably no more than a few million years, was required for the formation of asteroids from dust grains.

After the accretionary period, the evolutionary paths of the asteroids diverged. To a first approximation, asteroids nearer the Sun were heated more than those farther away because most heat sources increase in intensity closer to the Sun. For example, if bodies are heated mainly by electrical currents generated by the solar wind, bodies near the Sun will be more strongly heated because the Sun is the source of the solar wind. If bodies are heated mainly by impacts, bodies near the Sun will tend to be more intensely heated because impact velocities are greater near the Sun. And finally, if asteroidal heating is mainly caused by the decay of short-lived radioactive nuclei such as aluminum 26, asteroids nearer the Sun will get hotter because planetesimals accreted faster near the Sun and were thus able to incorporate greater amounts of undecayed nuclei.

Most asteroids, especially those in the outer parts of the asteroid belt, remained unmelted. They never segregated into bodies with iron cores and silicate mantles. Their entire geological history consists of occasional bombardment by smaller meteoroids, mild heating, and alteration by watery fluids. But some asteroids, particularly those concentrated in the inner part of the asteroid belt, did melt and differentiate. One of these was the mesosiderite parent asteroid.

The original, chondritic parent asteroid of the mesosiderites probably contained a fairly homogeneous distribution of silicates and metal. During melting, the metallic iron-nickel and silicates separated into immiscible liquids of different densities, much as a mixture of oil and water would after being shaken in a jar and left standing on the kitchen counter.

Scientists can model what happens when an entire primitive asteroid is melted: the denser metallic liquid sinks to the gravitational center of the body and forms a core. As the silicate melt above the core cools, crystals of olivine — a dense, dark green silicate mineral containing iron and magnesium, $(Fe,Mg)_2SiO_4$ — form and settle outside the core. Over the course of tens of millions of years, an olivine-rich mantle eventually covers the core. Heating of the mantle produces a low-density magma of basaltic composition that erupts to the surface and forms a basaltic crust consisting of fine-grained black rock. If the magma fails to make it to the surface, the liquid cools more slowly, and the crystals grow larger. A coarse-grained dark rock called a gabbro would form.

Once completely solidified, a typical differentiated asteroid would consist (by volume) of about 10–15 percent metallic iron-nickel core, 70 percent olivine-rich mantle, and 15–20 percent basaltic crust. If the asteroid had a radius of 100 kilometers, the radius of the core would be about 50 kilometers, and the basaltic crust would be about 5 kilometers thick.

This brings us to the central riddle of the mesosiderites. These meteorites seem to represent an extremely efficient mixing of crustal material (mainly basalt and gabbro) and core material (metallic iron-nickel and sulfide). If the basalts and the metal had formed in the same 100-kilometer body, they would have been initially separated by 45 kilometers of olivine-rich mantle. However, mesosiderites contain very little olivine, only about 1–2 percent. Where did it go?

Various models have been proposed over the years to account for the missing olivine. One involved the sinking of basaltic blocks of the crust through molten mantle to land atop the metal core. The problem was that the temperature of the molten mantle was above the melting temperature of the basaltic block. It should not have survived.

It now appears that the melting of the mesosiderites' parent body occurred shortly after the accretionary process. Isotopic studies of the silicates in mesosiderites suggest that melting took place a mere 40–140 million years after the formation of the oldest known refractory inclusions in chondritic meteorites.

Basalts that formed during the initial melting and differentiation of the mesosiderite parent asteroid constitute about 11 percent of the large silicate clasts in mesosiderites. These basalt clasts are indistinguishable in texture, mineralogy, and composition from the pure basalt meteorites known as eucrites. An isotopic analysis by Brian Stewart and colleagues at the California Institute of Technology indicates that a basaltic clast from the Vaca Muerta mesosiderite found in the Atacama Desert of Chile was formed 4.48 ± 0.09 billion years ago. This date is consistent with the rock having formed virtually simultaneously with accretion 4.56 billion years ago but does not preclude the possibility that it was formed up to 170 million years later. Isotopic ages need to be measured on additional basaltic clasts in mesosiderites to resolve this uncertainty.

We can also deduce some of the processes that formed the original basaltic crust of the mesosiderite asteroid. During the process known as fractional crystallization, crystals that formed from a cooling melt become isolated and are no longer able to react with the melt. A common way for crystals to become isolated is by settling to the floor of a magma chamber under the influence of gravity. Isolation of the crystals depletes the melt of the elements that are concentrated in the crystals. For example, the magnesium-iron silicate olivine is one of the first minerals to crystallize from liquids of basaltic composition; because early-formed olivine has a high magnesium/iron ratio and has less silicon than the melt, isolation of olivine from the surrounding liquid depletes the melt in magnesium more than it does in iron and enriches the melt in silicon. Crystals formed after the isolation of the olivine will therefore have lower ratios of magnesium to iron.

The geochemical characteristics of the basaltic clasts in mesosiderites indicate that they formed by fractional crystallization processes. This was the second major event in the history of the mesosiderite asteroid; it occurred within thirty million years of accretion and was the principal process that formed the asteroid's original basaltic crust.

The next major event in the evolutionary history of the mesosiderite parent asteroid is revealed by studies of the coarse-grained rocks called gabbros. These rocks constitute about 39 percent of the large silicate clasts in mesosiderites. The gabbroic clasts are similar in mineralogy to the basaltic clasts but have larger crystals.

As more and more crystals are formed in a magma chamber, the remaining liquid becomes enriched in those elements that cannot be readily incorporated into the crystal structures of the crystallizing minerals. These elements are called incompatible elements and are the last elements to go into crystals as the liquid is used up. In basaltic rocks they tend to be concentrated in glass or small, rare mineral phases in between the larger crystals of common minerals.

Most of the rare-earth elements, those with atomic numbers 57 (lanthanum) through 71 (lutetium), are among the incompatible elements. (Atomic numbers refer to the number of protons in the nucleus of an atom.) The rare-earth elements are large atoms that do not readily fit into the crystal structures of most minerals; hence, they tend to concentrate in residual liquids during crystallization.

The relative abundance of rare-earth elements in an igneous rock (i.e., a rock formed from a melt) provides a clue to the history of its crystallization. The basis of comparison is the relative abundance of rare-earth elements in chondrites; these meteorites have the same composition of rare-earth elements as the Sun (which is effectively the average composition of the solar system). Rocks with a composition of rare-earth elements matching those of chondrites have flat rare-earth element patterns (i.e., the abundances of the different rare-earth elements are all about the same when divided by the abundances of these elements in chondrites).

Geochemical analyses of gabbro clasts in mesosiderites indicate that these rocks have extremely low concentrations of most rare-earth elements (figure 9.2). Calculations indicate that the gabbros could not have formed by a single episode of crystal settling from a basaltic melt. Instead, they probably formed by a two-stage process involving crystal settling to form a rock depleted in incompatible elements, followed by remelting of this rock and the accumulation of crystals from this second-generation melt.

The large abundance of gabbro clasts in mesosiderites (and the two-stage heating mechanism required to form them) indicates that a major remelting of the crust took place. This is the third major event in mesosiderite history. Brian Stewart and coworkers determined the age of one of the gabbros to be 4.47 ± 0.15 billion years. They also dated a remelted basaltic clast at 4.51 ± 0.04 billion years. I provisionally take 4.5 billion years as the date of this event, but its absolute age is not well resolved from either accretion or differentiation.

At about the same time as the crust of the asteroid was remelting, there was some mixing of metals and silicates. Because most mesosiderites contain about half silicates and half metal, metal-silicate mixing is the main event in the formation of mesosiderites. It seems likely that the metal-silicate mixing event was the same event responsible for crustal remelting. During metal-

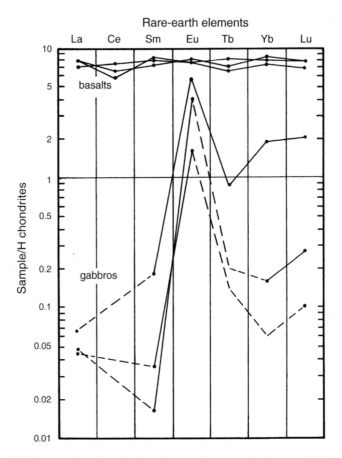

Figure 9.2. Diagram showing the abundance of rare-earth elements (REE) in meso-siderite clasts. All concentration data are divided by the mean REE concentrations in H-group ordinary chondrites. Basaltic clasts in mesosiderites, *top*, have flat REE patterns. In contrast, the gabbroic clasts in mesosiderites have highly fractionated patterns: they have very low abundances of the light REE (e.g., La, Ce, Sm), high abundances of Eu, and lower abundances of the heavy REE (e.g., Tb, Yb, Lu). Solid lines connect well-determined data points. Dashed lines connect estimated REE concentrations.

silicate mixing, phosphorus from metal was oxidized to produce the moderately abundant phosphate in the mesosiderite silicates.

Analyses of large metal nodules in fifteen mesosiderites demonstrate that their composition is similar to that of the metal in the largest group of iron meteorites — the so-called IIIAB group. (Iron meteorites can be chemically classified into about a dozen groups based on the abundance of germanium and gallium relative to that of nickel. The groups are labeled with a series of Roman numerals and letters.)

The IIIAB iron meteorites formed by fractional crystallization of metal in asteroid cores. Each of the two hundred or so known IIIAB meteorites represents the product of a different degree of crystallization and hence has its own distinct interelement ratios. For example, during fractional crystallization of metals, nickel has a slight preference for remaining in the liquid, whereas iridium is strongly partitioned into the solid — the result is that the iridium/nickel ratios in the earliest-formed IIIAB iron meteorites are approximately three thousand times greater than those of the latest-formed members of the group.

In contrast, mesosiderite metal is far more uniform in composition. The iridium/nickel ratios in the metal nodules we analyzed generally vary by less than a factor of two. Therefore, if the metal in mesosiderites was derived from a IIIAB-like core, then that core must still have been largely molten at the time metal-silicate mixing occurred because fractional crystallization could not yet have proceeded very far. This implies that the metal-silicate mixing event took place very early in the asteroid's history because the core would have become fully crystallized about one hundred million years after it was formed. This conclusion is consistent with the approximate date of four and a half billion years assigned to this event based on isotopic dating.

The simplest model for mixing metal and silicate is a collision between two differentiated asteroids. The reason this model had not gained wide acceptance in the past is that most terrestrial and lunar impacts suggest that projectile material is widely dispersed over the surface of the target asteroid. The soils at several lunar landing sites suggest that the ratio may be as little as one part projectile to one hundred parts of the target body. The absence of mesosiderite analogs on Earth and the Moon (even from craters known to have been formed by impacting iron meteorites, such as Meteor Crater in Arizona) is the result of the high relative velocities of impacting projectiles. During such collisions the projectile material is largely vaporized, and much of the remainder is diluted by vastly larger amounts of target ejecta.

In the present-day asteroid belt, bodies collide at velocities of about 5 kilometers per second, too high to allow the projectile material to remain largely undispersed. However, in the early solar system, during and shortly after accretion, there were many bodies in low-inclination, nearly circular orbits. (If this had not been the case, the collisional velocities among the asteroids would have been too high for accretion to have occurred.) In this early period, the average collisional velocity may have been less than 1 kilometer per second. It is precisely during this period that cores of differentiated asteroids were still largely molten. It therefore seems plausible that the basaltic surface of the mesosiderite parent asteroid was struck at a low velocity by an asteroid having a largely molten core (figure 9.3).

Heat generated by this collision may have been responsible for remelting the crust of the asteroid and producing gabbroic rocks with very low abundances of rare-earth elements. The rare coarse olivine grains in mesosiderites may have been acquired from fragments of the mantle overlying the core of the impacting asteroid. No comparable event involving metal-silicate mixing, oxidation-reduction, or crustal remelting is evident among the eucrites, suggesting that they were derived from a separate parent asteroid.

The fourth major epoch in the history of the mesosiderite asteroid was characterized by localized impact melting that occurred between 4.5 and 3.9 billion years ago. Small blobs of melt were produced during these small-scale impact events; the blobs cooled quickly into fine-grained rocks. Brian Stewart and his colleagues found that one of these fine-grained impact-melted rocks from the Vaca Muerta mesosiderite has an age of 4.42 ± 0.02 billion

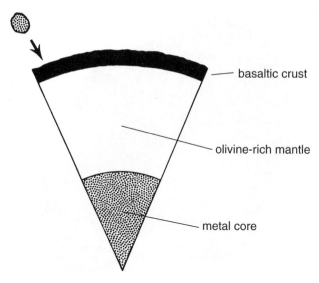

Figure 9.3. A simple model for mesosiderite formation featuring the low-velocity impact of a metallic projectile with the basaltic crust of a differentiated asteroid.

years. Other researchers found silicates in the Estherville, Iowa, mesosiderite that range in age from 4.5 to 3.9 billion years. Most of this material probably represents fragments of impact-melted rocks produced by many different impact events during this period.

These impact events dramatically affected the textures of individual mesosiderites. Some of these meteorites experienced relatively little recrystallization (figure 9.4), some were moderately recrystallized, and some have been severely affected by impact melting (figure 9.5).

The fifth, and most recent, epoch of mesosiderite history involves a series of processes that happened long after the parent body cooled. The mesosiderites were excavated from their burial sites by meteoroid collisions and deposited on the surface of the asteroid. Noble-gas measurements indicate that the Veramin mesosiderite acquired solar-wind gases at this time. Noble gases such as helium, neon, argon, and krypton are carried through interplanetary space by the solar wind; when they encounter po-

Figure 9.4. Slab of the Mount Padbury mesosiderite. In addition to the dark pyroxene-rich clasts and white metal-sulfide nodules, there are several individual metal-rich mesosiderite fragments embedded in the host. (Courtesy of the Smithsonian Institution)

rous rocks on airless bodies such as the Moon and the asteroids, some of the gas molecules attach themselves to the surfaces of grains. Enrichment of a rock in solar-wind gases is *prima facie* evidence that it was once exposed at the surface of an airless body.

Some time after their excavation, the mesosiderites were ejected from the surface of their parent asteroid and launched into interplanetary space by a series of impact events. While in space, the mesosiderites were bombarded with galactic cosmic rays — high-energy protons originating outside the solar system. These particles are energetic enough to cause nuclear reactions in the materials they penetrate, producing both radioactive and stable isotopes. Measurements of the ratios of the cosmic-ray-produced radionuclides and their daughter isotopes allow the determination of the cosmic-ray exposure age of a meteorite. This is actu-

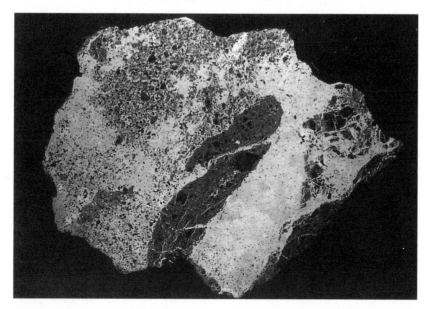

Figure 9.5. Slab of the anomalous mesosiderite, Reckling Peak A79015. There are elongated and irregular pyroxene-rich clasts embedded in a host that contains highly variable proportions of metal and silicate. The slab is about 15 cm across. (Courtesy of the Smithsonian Institution)

ally a measure of the duration of the meteorite's residence in interplanetary space as a meter-size (or smaller) object. The cosmic-ray exposure ages of mesosiderites range from 10 million to 160 million years. The mesosiderites' journey through interplanetary space was eventually cut short by yet another collision — this time with Earth.

The most massive mesosiderite known to have struck Earth was a 0.5-kilometer-diameter asteroid that impacted the South Pacific Ocean 2.15 million years ago. Tiny fragments of this asteroid were discovered by UCLA geochemist Frank Kyte in deep-sea sediment cores recovered hundreds of kilometers apart. Because the pieces were recovered near the Eltanin Fracture Zone on the Pacific Ocean floor, the meteorite was named Eltanin. Analyses of pieces of this large mesosiderite indicate lower levels of siderophile elements than other mesosiderites. This suggests that metal mixing was inhomogeneous on the mesosiderite parent body and that small, metal-rich mesosiderites may be unrepresentatively rich in metal.

Future work will undoubtedly modify our inferences about the history of the mesosiderite parent body. In particular, additional isotopic analyses will refine the chronology of events. In the meantime, the scenario depicted here illustrates the complicated geological history of asteroidal bodies. Recognition that collisions have played an important role in their history comes on the heels of the realization that Earth has been bombarded by enormous projectiles as well. Some of these projectiles, such as Eltanin, have been mesosiderites; some, such as the one responsible for the major extinction event that wiped out the dinosaurs, have been carbonaceous chondrites; and some have been iron meteorites. In the next chapter we explore the product of the collision of one of these iron meteorites — Meteor Crater, Arizona.

X

METEOR CRATER

O

> With a sudden crash came thunder on the left,
> and a shooting star trailing a firebrand
> slid from the sky through the dark
> and darted downwards in brilliant light. . . .
> It left a long luminous streak in its wake,
> and far around a sulphur-smoke was seen to rise.
> —Virgil, *Aeneid*

METEOR CRATER is the freshest impact crater on Earth (figure 10.1). It is a quasi-square-shaped depression, 1.2 kilometers from rim to rim, 4.8 kilometers in perimeter, and about 200 meters deep. In comparison, the height of the Washington Monument is 169 meters, while that of the Great Pyramid of Khufu is 137 meters.

The crater lies in the northern Arizona Plateau about halfway between the cities of Winslow and Flagstaff. The area is semiarid with outcroppings of Triassic-aged Moenkopi sandstone and Permian-aged Kaibab limestone. Evidence of volcanism is everywhere. Buttes and cinder cones dot the Painted Desert along the Colorado River. Within this desert lies the Petrified Forest—trees

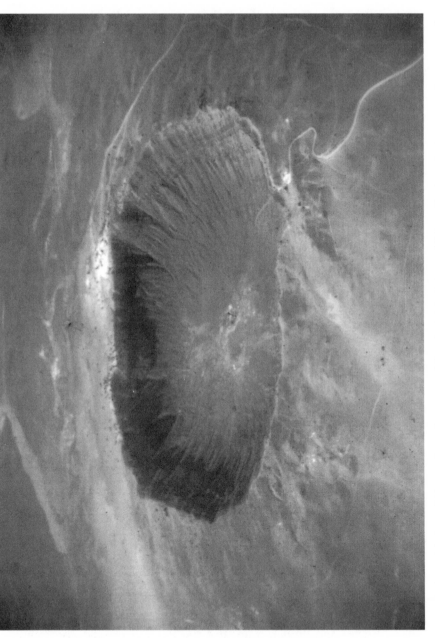

Figure 10.1. Aerial view of Meteor Crater. Remnants of the ejecta blanket are visible as the irregular, bumpy surface surrounding the crater rim. (Courtesy of Edward J. Sparling, Roosevelt University)

buried by volcanic ash in Jurassic times. This is a quiet, ancient land, very dry and unmarred by glaciers. The crater could not have been so well preserved in a more temperate climate.

The first white settlers arrived in this region around 1871. They were cowherds and had little time to explore "Coon Butte" or "Coon Mountain" as the crater was then called. A man would have had to travel several hours on horseback, facing gusts of wind of up to 100 kilometers per hour, in order to reach the crater from the nearest settlement. Most people assumed that the crater was just another extinct volcano, and so spent their time more practically, tending to the needs of their livestock.

In 1876 a large chunk of iron was picked up near Canyon Diablo, about 3 kilometers west of the crater, by a cattleman named Mathias Armijo. Thinking his find was iron ore, he tried to sell it.

Prospectors later came to the crater and were quite impressed by the amount of iron fragments scattered over the terrain (figure 10.2). Samples of this iron eventually were sent to the mining firm of M. W. Booth & Co. in Albuquerque, New Mexico. In April of 1891 an assayer in Denver, Colorado, examined a piece of the Coon Butte iron and reported finding a composition of 76.8 percent iron and 1.8 percent lead, with traces of gold and silver. This analysis, improperly conducted by an inexperienced assayer, showed a very unmeteoritic composition for the specimens. (Modern analyses reveal an average content in weight percent of 92.1% iron, 6.9% nickel, 0.47% cobalt, 0.26% phosphorus, 0.14% carbon, 0.03% germanium, 0.01% copper, 0.01% sulfur, 0.01% gallium, and trace amounts of other metals.)

The mining firm sent other samples elsewhere for examination. One of the irons was shown to Dr. Albert E. Foote, a physician, former professor of chemistry and mineralogy at the University of Michigan, and, at the time, a prominent mineral dealer from Philadelphia, Pennsylvania. Foote immediately recognized the substance as meteoritic. Intrigued by the report that "a carload of similar material could be gathered from the surface," Foote visited the Arizona crater in June 1891 (figure 10.3).

Figure 10.2. A 15.9-kg specimen of the Canyon Diablo iron meteorite. This is a fragment of the projectile that formed Meteor Crater. Ruler at the bottom is approximately 15 cm in length. (Courtesy of Blaine Reed)

Foote examined the area thoroughly and noted that he could find no traces of obsidian, lava, or any of the other usual products of volcanic activity. He and his assistants collected 137 meteorite fragments, each weighing between a few grams and 3 kilograms. These meteorites had occupied an oblong region up to 36 meters wide, stretching 518 meters from the crater rim to the southeast. Foote also recovered two large irons (weighing 70 and 91 kilograms) along the axis of this region about 3 kilometers from the crater's southeastern rim.

In the published account of his findings, Foote attempted no explanation of the crater's origin, merely labeling it "a remarkable geological phenomenon."

Further analysis of the iron meteorites at Pennsylvania University indicated small inclusions of diamond. Even though these

Figure 10.3. View of Meteor Crater from just below the rim.

specks were of no commercial value, their existence drew far more immediate interest than the crater.

Foote's paper on the Canyon Diablo meteorites was read by Dr. Grove K. Gilbert, chief geologist of the U.S. Geological Survey. Only four years later, Gilbert was to publish his bold proposal that lunar craters resulted from meteorite impact instead of from volcanic activity. Gilbert was busy with bureaucratic duties and dispatched Willard D. Johnson, a civil engineer and topographer, to examine the crater. As Gilbert later paraphrased Johnson's report, "in some way, probably by volcanic heat, a body of steam was produced at a depth of some hundreds or thousands of feet and the explosion of this steam, produced the crater."

This report initially seemed quite unsatisfactory to Gilbert, so he set out in the fall of 1891 to investigate the crater along with colleague Marcus Baker, an expert in measuring terrestrial magnetism. Extensive measurements were made, a topographical map was drawn, and numerous photographs were taken. More meteorites were unearthed and critically examined. The lack of volcanic rocks in the crater seemed to point to the impact hypothesis, but the absence of magnetic anomalies pointed away. Gilbert was sure that if a large mass of iron was buried beneath the crater floor, it should have been detected by Baker's magnetic instruments. Gilbert did not realize that the enormous energy evolved during the impact had vaporized the projectile. He turned his attention to the possibility of a colossal steam explosion. Gilbert had been impressed by the tremendously energetic explosion just eight years earlier of the Krakatoa volcano in the strait between Java and Sumatra. The sound of that explosion was heard 2,000 km away in Australia. Gilbert reasoned that Krakatoa must have released an enormous amount of steam and thought it possible that a deep-seated, Krakatoalike steam explosion could have formed the Arizona crater. In the end, Gilbert admitted that Johnson's hypothesis was probably correct—the crater was of volcanic rather than of meteoritic origin. It was taken as mere coincidence that more iron meteorites had been found in the vicinity of the crater than the total that had been recovered everywhere else on Earth.

Gilbert was so highly regarded by his fellow geologists that after this pronouncement, investigations of the Arizona crater came to a virtual halt. For several years afterward the only papers even mentioning the Canyon Diablo irons dealt exclusively with their microscopic inclusions of diamond.

In October 1902 a Philadelphia mining engineer named Daniel Moreau Barringer heard about the crater and the incredible abundance of iron meteorites in the immediate vicinity. He was initially skeptical of the meteoritic origin of the crater that was insisted upon by some of the local inhabitants, but his interest was aroused. Several months later, Barringer and a mathematician and physicist named Benjamin C. Tilghman investigated the crater and "collected an astounding array of evidence in favor of [its impact origin]." They became firmly convinced that the main mass of the giant meteorite was still buried somewhere beneath the crater floor. On behalf of their newly formed Standard Iron Company, they laid claim to the crater under the U.S. Mineral Land Laws.

Barringer believed that since the crater was approximately circular, the meteorite must have fallen vertically; and so in 1905, when excavations were begun, twenty-eight holes were drilled in the center of the crater floor. Mine shafts were also sunk, on the rim as well as in the crater.

Tilghman and Barringer were impressed by the similarity of this crater with those formed by the explosions of large projectiles (e.g., cannonballs) traveling at high velocities. They stated (in complete agreement with Foote's analysis) that "in and around this hole and below its bottom to a distance of over 1,400 feet below the present surface of the plain surrounding it, and the original surface of the place where this hole was formed, every indication of either volcanic or hot spring action is positively absent." Many more meteorite fragments were found, some lying under immense boulders of Kaibab limestone apparently hurled

Figure 10.4. Remains of a hoist engine and steam boiler on the crater floor. This equipment was used by Daniel Barringer in early attempts to locate a hypothesized large iron meteorite block buried below the surface.

from inside the crater. Great quantities of pulverized Coconino sandstone that could be produced by a violent explosion but not by natural erosion were discovered. Meteoritic iron had been found in the borings at a depth of 150 meters below the present crater floor (figure 10.4). It was also found that about 400 meters beneath the surface of the original plain, undisturbed strata of Supai sandstone were present. If the crater had been formed by volcanic heat, as Johnson had suggested in the early 1890s, a great body of steam would have had to have originated "at a depth of some hundreds or thousands of feet," thus making it impossible to find strata of undisturbed rock beneath the crater.

"In view of these positively established facts," Tilghman reported in 1905, "the author feels that he is justified, under due reserve as to subsequently developed facts, in announcing that the formation at this locality is due to the impact of a meteor of enormous and hitherto unprecedented size."

After these investigations, Barringer began to hunt for the main mass of the huge meteorite with renewed vigor. The main shaft could not be sunk more than 68 meters below the crater floor, because of the water-saturated silica encountered there. The drill holes had bored to far greater depths, however, without locating any large masses of iron.

Barringer noted that when a rifle bullet was fired into thick mud, it always made a round hole (never elliptical) even at obtuse angles of incidence. This immediately suggested that the meteorite also may have arrived from a great angle. Foote's earlier description of the elliptical region of irons stretching from the base of the crater's northwestern rim to the southeast seemed to indicate that the original meteorite came from a northerly direction. The occurrence of many small oxidized meteorites, which Barringer referred to as "shale-balls," on the northern rim also implied that the meteorite had come from the north. Yet another piece of evidence was uncovered—the rock strata in the northern portion of the crater were inclined only 5° to the horizontal, but in the southern portion these strata were almost vertical.

The picture now seemed somewhat clearer. The meteorite had

evidently approached from slightly west of north. It was therefore decided to dig under the crater's southern rim. A 6-inch (15-centimeter) churn drill was employed, and below 300 meters increasing amounts of meteoritic material were encountered. The going was extremely rough, however, often taking an entire day's drilling to advance less than half a meter.

In 1908 the well-known meteoriticist Dr. George P. Merrill visited the Arizona crater and consulted with Barringer about previous investigations. Merrill's published account indicates that he also was largely convinced about the crater's meteoritic origin. He imagined that the original mass must have been about 150 meters in diameter, traveling at a speed of 8 kilometers per second. Merrill concluded that the kinetic energy of such an impact would have destroyed the meteorite almost completely. Barringer's failure to find the main mass of the meteorite, even under the southern rim, attested to Merrill's conclusion.

Despite the fact that the investigations of Barringer and Merrill had uncovered strong evidence for the meteoritic origin of the crater, the U.S. Geological Survey tenaciously clung to the volcanism hypothesis. This made it difficult for Barringer to obtain more funds for continued drilling. In addition, of those astronomers who did accept the Arizona site as a true impact feature, many felt (as Merrill had) that the bulk of the meteorite must have been instantly vaporized. If so, further drilling could no longer be considered a viable commercial venture.

Barringer remained unruffled, however, and still vigorously insisted that if only more funds could be secured, the main block of the meteorite (which he estimated to weigh at least several million tons) could be located. He calculated that the amount of iron ore that might be mined from such a mass would be worth $250 million to the stockholders of his company. But the science of meteoritics was in its infancy, and no large meteorite craters were then known; thus, acceptance of Barringer's ideas was slow in coming. As late as 1919, a veteran geologist and mining engineer, Hugh M. Roberts, studied Coon Butte and the surrounding area, only to choose volcanism as the probable mechanism of the crater's formation.

Finally, in 1920 Barringer's company acquired fresh funds, and drilling commenced on the southern wall of the crater. Three years later operations ceased when the bore became wedged and finally lost. Barringer was convinced that he had at last struck the main mass, located about 430 meters below the southern rim. Unfortunately the Standard Iron Company had "greatly exceeded the estimated cost of drilling the hole" and refused further financing.

Barringer had found that the water level beneath the crater floor was encountered at a depth of 60 meters, still 100 meters or so above the level where he believed the main mass was located. Assuming that the water was limited to the crater itself, Barringer sank a shaft 300 meters from the southern rim when funds again became available. At a depth of 230 meters, however, water began pouring in at such a rate that three days of pumping were required for each day's drilling. Work had to be abandoned.

After some $250,000 had been spent investigating the crater, Dr. Forest R. Moulton was consulted to see if he could estimate just how much iron might be residing in the crater awaiting any future mining attempts. Moulton was an expert astronomer, highly versed in celestial mechanics, mathematics, and ballistics. Without journeying to the crater himself and relying solely on the data secured by Barringer, Foote, and Merrill, Moulton turned in his first report on 24 August 1929. It appeared to him that the meteorite had entered Earth's atmosphere as a compact cluster between 120 and 900 meters in diameter. Its mass was between 50,000 and 3 million metric tons. The swarm would have been traveling at cosmic velocity, 11 to 24 kilometers per second, and would have possessed such enormous momentum that air resistance could not have slowed the meteorite down.

Two years later in his textbook, *Astronomy*, Moulton wrote: "The energy given up in a tenth of a second would be sufficient to vaporize both the meteorite and the material it encountered—

there would be a violent explosion that would produce a circular crater, regardless of the direction of impact, which alone would remain as evidence of the event."

Moulton's report was accepted by the majority of scientists, who by this time had come to regard the Arizona crater as a true impact site. But Barringer's stockholders were horrified. There was apparently no main meteoritic block to be mined. Many letters were quickly exchanged between Moulton and Barringer and Barringer and the stockholders. Moulton reaffirmed that the meteorite must have been vaporized on impact, but Barringer would not be dissuaded. Unfortunately, the mining company's funds had been exhausted, and no material gain had been realized from the entire venture. Barringer's health rapidly deteriorated, and he died at his home in suburban Philadelphia.

Seventeen years after Barringer's death in 1929, the name Meteor Crater was officially adopted by the U.S. Department of Interior's Board of Geographical Names.

Other studies of the crater were later undertaken, many based on modern geophysical methods of investigation. In 1931 J. J. Jakosky, a consulting engineer for International Geophysics, Inc., conducted magnetic, geological, and electrical surveys of the crater. His results indicated that the crater might have formed as long ago as fifty thousand years. The presence of a large mass of foreign material buried under the southern part of the crater was also reported. In accordance with Jakosky's suggestions, two holes were drilled through the crater's southwestern floor, but these failed to turn up significant amounts of meteoritic material.

Numerous magnetic and gravimetric analyses of the crater were made through the 1950s; some indicated the presence of a meteorite block, others exactly the opposite. The data are complex and difficult to interpret. The anomalously high density reported in several distinct regions of the crater is probably a result of compression of porous Coconino sandstone by the tremendous pressure generated by the meteorite's impact.

After 1909, Barringer had ceased his systematic field investigations and had instead devoted all his time prospecting the crater for the missing iron block. Thirty years passed before Harvey H. Nininger, concerned about the dearth of field data, began his extensive investigations. Since numerous individuals had searched the crater and surrounding area for the larger meteorites and had, for the most part, kept no accurate records of their finds, Nininger thought it best to hunt for the very small fragments. These, at least, would have escaped previous detection. After raising some funds and constructing a magnetic rake of his own design, Nininger combed the north, northwest, west, and southwest sides of the crater. Investigations were conducted on the other portions of the crater to a much more limited extent because of the rugged terrain.

Altogether about twelve thousand small fragments were collected from the area, possessing a total weight of 19 kilograms. The distribution of these irons suggested that they had been thrown from the crater with great force.

In a footnote to his 1956 book, *Arizona's Meteorite Crater*, Nininger predicted that the recently described high-pressure, compact form of silica called coesite would have been formed in the sandstone on the crater floor during the impact. This phase was found four years later by Edward C. T. Chao in rocks from Meteor Crater; by 1962 Chao and coworkers had found stishovite, the highest-pressure form of silica, intermixed with coesite at Meteor Crater.

In 1956 John S. Rinehart from the Smithsonian Astrophysical Observatory investigated pulverized material in the soil around Meteor Crater. Seven hundred soil samples were analyzed, and three magnets were used to separate the strongly magnetic meteoritic material from slightly magnetic terrestrial material. Inevitably, shiny black particles of terrestrial magnetite appeared in the final sample, but they were easily identified and discarded. From the amount of pulverized nickeliferous iron in the samples, the total quantity of meteoritic material in the soil around the crater was calculated to be about 10,000 metric tons.

Sedimentary material gathered from mine shafts in the crater floor by Eugene M. Shoemaker between 1959 and 1962 included a few freshwater shells deposited in the crater shortly after its formation. Carbon-14 tests indicated that these shells were about 22,000 years old, yielding a minimum estimate of the crater's age. Steve R. Sutton made thermoluminescence measurements on shocked quartz grains from sandstone on the crater floor in 1985 and found an age of 49,000 ± 3,000 years for the impact event that formed the crater. In 1991 Kuni Nishiizumi and coworkers measured the beryllium 10 and aluminum 26 isotopic exposure ages of rocks on the crater walls and on top of large blocks on the crater rim. They determined a minimum age of 49,200 ± 1,700 years for Meteor Crater. This age is in accord with other recent independent measurements and is the age accepted today.

Shoemaker made the first comprehensive geologic studies of Meteor Crater. He mapped the crater, noting overturned blocks of sandstone at the crater rim and blocks of ejecta in the surrounding plains. He discovered a 150-meter-thick lens of brecciated rocks (rocks composed of jumbled and broken rock and mineral fragments fused together by impact) beneath the floor of Meteor Crater.

Shoemaker longed to go to the Moon as the first geologist-astronaut, but a 1963 diagnosis of Addison's disease grounded him permanently. Since he could not go himself, Shoemaker lobbied hard to get other scientists involved in the space program. He also trained many of the Apollo astronauts in crater mechanics and sampling techniques at Meteor Crater. Here, the similar morphologies of lunar and terrestrial impact structures helped prepare the astronauts for their missions. When *Apollo 16* astronauts John Young and Charlie Duke toured the Cayley Plains in the Lunar Rover in 1972, they found brecciated rocks everywhere similar in structure to some of the rocks at Meteor Crater.

Shoemaker and other researchers were able to infer the processes that formed Meteor Crater. The crater was produced by the impact of a 50-meter-diameter metallic iron projectile that liberated an amount of energy equivalent to 20–60 megatons of TNT. The explosion excavated a depression 150 meters below the surrounding plains and formed a rim 67 meters high. More than 100 million tons of rock were thrown from the crater. Debris slid down the slopes at high speeds and piled up into a 15-meter-high mound on the crater floor. Subsequent erosion lowered the rim by 20 meters; sedimentation filled in a substantial fraction of the excavated cavity.

At the instant of impact, a shock wave was produced that raced through both the target and the projectile. The mass of the projectile became highly compressed as the shock overcame the material strength of the object. When the shock wave hit the rear surface of the projectile, a small fraction of the rear surface was ejected and survived as lightly shocked material. The shock wave itself (with only slightly diminished energy) was reflected back through the compressed projectile as a rarefaction wave toward the ground. This type of wave traveled through the compressed projectile at the speed of sound and dramatically reduced the pressure in the object. At this point, most of the projectile vaporized and formed an iron vapor plume rising out of the crater. Some of the vapor condensed into small iron spherules and rained out over the countryside.

Geologist David Kring pointed out that a powerful air blast was caused by the shock wave, scouring the landscape with wind speeds exceeding 1,000 kilometers per hour. Trees and grasses were uprooted, and the Ice Age animals within a few kilometers of the crater were killed either by the air blast itself or by being pelted with branches, stones, and sand.

As the freshest impact crater on Earth, Meteor Crater has proven invaluable in gaining understanding of solar-system impact processes. But it is not the only fresh crater visible from Earth. At the same time that scientists were struggling to identify the mechanism that produced Meteor Crater, they were also trying to figure out how lunar craters formed. The recognition that lunar craters were formed by impacts is the story of the next chapter.

XI

THE LUNAR CRATER CONTROVERSY— A BRIEF RETELLING

And as the smart ship grew
In stature, grace, and hue,
In shadowy silent distance grew the Iceberg too.

Alien they seemed to be:
No mortal eye could see
The intimate welding of their later history,

Or sign that they were bent
By paths coincident
On being anon twin halves of one august event,

Till the Spinner of the Years
Said "Now!" And each one hears,
And consummation comes, and jars two hemispheres.
—Thomas Hardy, *The Convergence of the Twain*

IN 1609, the same year that Shakespeare penned the tragicomedy *Cymbeline*, replete with circuitous plot twists, Galileo trained his 1.5-inch (3.8-centimeter) telescope at the Moon and saw numerous circular spots. During the course of the long lunar day, Galileo observed that the Sun lit the rims of the spots

Figure 11.1. The lunar crater Eratosthenes (58 km in diameter) near the edge of Mare Imbrium, photographed by the *Apollo 13* astronauts. (Courtesy of NASA)

before lighting their floors and concluded that the spots were in fact depressions, that is, craters. He noted that many of the craters had central peaks and many had dark-colored floors (figure 11.1).

After Galileo, more-detailed maps of the Moon were made by other astronomers. A 1647 map by Hevelius shows numerous craters including Tycho, with its prominent system of rays.

The origin of lunar craters was first addressed by the English physicist and chemist Robert Hooke in his 1665 book *Micrographia*. Hooke described a series of experiments wherein he attempted to duplicate lunar-crater morphology. He dropped musket balls and clumps of mud into a clay-water slurry and found that the resulting pits resembled lunar craters. However, because the prevailing scientific view was that interplanetary space was empty, Hooke rejected the impact hypothesis. He boiled up some alabaster, inspected the pits on the surface and noted their similarity to lunar craters. He concluded that lunar craters were volcanic landforms.

Hooke's view remained essentially unchallenged for more than a century. In 1785 the German philosopher Immanuel Kant came out in favor of lunar volcanism, and two years later the great observational astronomer William Herschel reported a volcanic eruption on the Moon. In 1791 Johann H. Schröter published the results of his seven-year telescopic investigation of the Moon. He inferred the heights of lunar features by the lengths of their shadows, compared lunar and terrestrial landforms, and concluded that volcanoes must have formed the lunar craters.

Progress toward an understanding of lunar craters stalled until near the end of the eighteenth century. During a thirteen-year period, between 1794 and 1807, three series of events combined to make the impact theory more plausible. First, the German physicist Ernst F. Chladni published a monograph in 1794 demonstrating the existence of meteorites. He linked the common observations of fireballs to the much rarer reports of rocks falling from the sky. Second, the first asteroid, Ceres, was discovered in 1801 orbiting the Sun between Mars and Jupiter. Three more asteroids were discovered by 1807. Third, a series of well-documented meteorite showers occurred in Europe, India, and America between 1794 and 1807, forcing many previously skeptical scientists to concede that extraterrestrial rocks do sometimes fall to Earth. Interplanetary space no longer seemed so empty.

The idea of an impact origin for lunar craters suffered a setback in 1829 when it was endorsed by Franz von Paula Gruithuisen. Seven years earlier Gruithuisen had reported the existence of men and animals on the Moon, thereby earning the opprobrium of his scientific colleagues. Gruithuisen's acceptance of the impact hypothesis caused some mainstream astronomers to shy away from it. Although the impact idea was mentioned occasionally in the nineteenth century, the volcanic hypothesis was widely accepted. Detailed maps were made of the lunar surface by astronomers who strongly supported volcanism. The well-known American geologist James D. Dana visited Hawaii, trekked across the shield volcanoes, and compared lunar craters to volcanic calderas (large explosion or collapse craters generally near the volcano summit). Dana learned that calderas could become enlarged by explosion or collapse and suggested in 1846 that lunar craters could have formed by an analogous process.

The volcanic hypothesis received significant public attention when Jules Verne discussed it in his 1869 novel *All around the Moon*. Verne described the lunar crater Copernicus as containing numerous volcanic cones, each $\frac{1}{2}$ mile (800 meters) high. The bright rays radiating from the crater were identified with shrinkage cracks or cooling cracks through which lava was forced onto the lunar surface by internal pressure. From there it oozed over the plains and flooded the landscape.

Breaking with tradition, Grove K. Gilbert of the U.S. Geological Survey provided major support for the impact hypothesis in 1893. His influential article "The Moon's Face: A Study of the Origin of Its Features" appeared in the *Bulletin of the Philosophical Society of Washington*. It provided depth/diameter ratios for lunar craters, correctly explained central peaks as due to the rebound of viscous target material, crater rays as ejecta, and terraces within craters as slumped crater walls. He showed that circular depressions analogous to lunar craters could be created by low-velocity impacts of projectiles into slurries, powders, and porridge.

But Gilbert faced a major problem—his low-velocity experiments produced circular craters only when the impacts were vertical. Oblique impacts produced elliptical craters. Because almost all lunar craters are circular, Gilbert's experiments implied that nearly all impacts onto the lunar surface were vertical. Gilbert proposed that the Moon had been struck mainly by bodies in

Earth orbit that fell vertically onto the lunar surface. Astronomers realized, however, that projectiles could hit the Moon at virtually any angle, and by and large rejected the impact theory.

The solution to Gilbert's dilemma was slow in coming. In his investigations of Meteor Crater, Arizona, Daniel M. Barringer noted circa 1905 that rifle bullets fired into sediment left a circular hole even at low angles of entry. Barringer realized that the reasonably round outline of Meteor Crater did not require a vertical impact; a *high-velocity* impact could produce a circular crater at any angle except very oblique ones (say, less than 20°). Ironically, Barringer's solution did not catch on among contemporary scientists in part because Gilbert himself had concluded in 1891 that Meteor Crater was volcanic.

Other investigators independently hit upon Barringer's solution. The Irish-Estonian astronomer Ernst J. Öpik found in 1916 that high-velocity impacts produced circular craters at nonoblique angles of incidence. But Öpik's finding was overlooked by Western scientists because it was published in the middle of World War I in Russian in an obscure Estonian journal. In 1919 American physicist Herbert E. Ives compared high-velocity impacts to exploding bombs, noted that many bomb craters have central peaks, and concluded that "the shape of the cavity has no reference to the angle at which the bomb strikes, but takes its form from the symmetrical explosive forces." Although he published his article in the mainstream *Astrophysical Journal*, Ives's conclusions failed to persuade most astronomers and brought strong criticism from William W. Campbell, the director of Lick Observatory.

An oft-repeated objection to the impact hypothesis was that if lunar impact craters were so abundant, then there ought to be a similar number on Earth. There obviously is not. Even if Meteor Crater was accepted as an impact feature (and this view was not widely held until the 1930s), it seemed to be the only such feature on Earth. In 1921 Alfred Wegener, the German meteorologist and early proponent of continental drift, supported the impact origin of lunar craters but dodged the problem of the apparent paucity of terrestrial impact craters. He stated that "the absence of large impact craters on the Earth can in no way be used against the hypothesis of lunar impacts."

Some scientists faced the problem head-on. A few conjectured that Earth's atmosphere served as an effective shield protecting the surface from the impacts of large projectiles (figure 11.2). (We will see later that, although this may be true for many projectiles tens of meters in radius, kilometer-size objects generally pass through the atmosphere without disruption.) In 1937 Leonard J. Spencer realized that erosion would have destroyed ancient craters on Earth. He concluded that old craters would be "destroyed by weathering and all traces of them obliterated by denudation long before the next geological period."

The same conclusion was reached independently by Ralph B. Baldwin in 1949. Baldwin strongly advocated the impact origin of lunar craters and realized that although erosion on Earth had indeed destroyed most terrestrial craters, their structural remnants might still exist. He concluded that "written in the Book of Geology in still obscure characters are the records of hundreds of thousands of collisions of the earth and extraterrestrial bodies." Additional support for the impact model of lunar craters came from American geologist Robert S. Dietz in 1946 and Baldwin's 1949 observation that the depth/diameter ratios of bomb craters, terrestrial impact craters, and lunar craters were about the same (figure 11.3).

In 1952 Nobel laureate Harold C. Urey published his highly influential book *The Planets* and essentially founded the modern field of planetary science. Urey advocated an impact origin for lunar craters and convinced many astronomers and geologists. Two years later, astronomer Gerard P. Kuiper published a widely read paper, "On the Origin of Lunar Surface Features," that inferred support for the impact model from photographs of lunar craters taken with large telescopes. More astronomers were convinced.

As more terrestrial craters were identified in the 1950s and 1960s and found to have depth/diameter ratios fitting Baldwin's

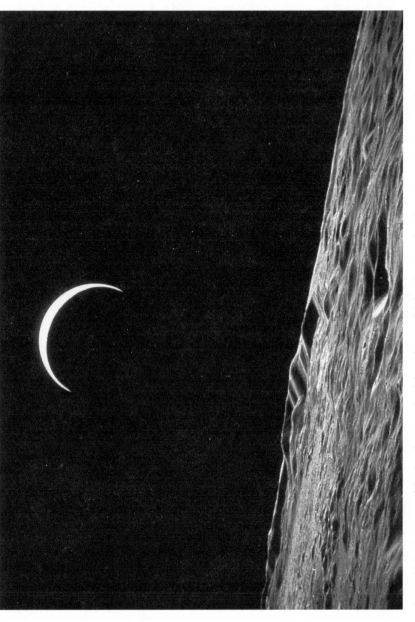

Figure 11.2. Crescent of Earth, photographed from lunar orbit by the *Apollo 15* astronauts. (Courtesy of NASA)

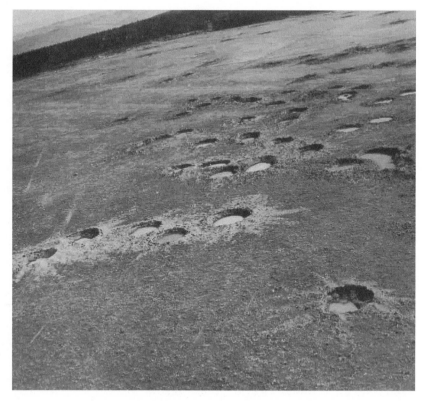

Figure 11.3. Low aerial photograph of bomb craters on Lippe Airfield in western Germany. This base was attacked by Allied bombers (dropping medium-size bombs) during World War II on 13 March 1945. The attacks were designed to prevent the remaining planes of the Luftwaffe from taking off. The craters display ejecta blankets and rays resembling those of fresh lunar craters. The bomb craters and lunar impact craters have very similar depth/diameter ratios. (Courtesy of Air Force Historical Research Agency, Maxwell Air Force Base, Alabama)

curve, the case for impacts became even stronger. The identification of shock features and high-pressure minerals in some apparently nonvolcanic terrestrial craters proved their impact origin. Similarly, after the return of the Apollo samples, lunar scientists identified shock features in lunar minerals and reported the occurrence of impact glass, chondrulelike crystalline impact-melt spherules, and hypervelocity micrometeorite craters in lunar samples. They examined a wide variety of brecciated moon rocks and concluded that countless impacts have modified the lunar surface at all size scales.

Of course, volcanic features occur on the surface of the Moon, a fact that lunar researchers have never denied. The lunar maria are vast plains of flood basalt, the dark-colored floors of some lunar impact craters are composed of the same sort of basalt, the Marius Hills appear to be small volcanic domes, and sinuous rilles (narrow meandering trenches) on the maria seem to be collapsed lava tubes. However, because volcanism ceased more than two billion years ago, William Herschel's 1787 report and hundreds of others by later, less gifted observers of erupting lunar volcanoes are apparently mistaken. Bad optics, terrestrial atmospheric turbulence, and overactive imaginations are presumably responsible for these phantom lunar volcanic eruptions.

Today's lunar scientists are comfortable with the concept that the Moon is a geologically complicated body whose surface has been shaped by the impacts of comets, asteroids, meteoroids, and micrometeoroids as well as by volcanism, gravitational slumping, and solar-wind bombardment. But it was the recognition that lunar craters are impact features that paved the way for a more complete understanding of the geological history of the Moon. This understanding also allowed planetary scientists to get a handle on the early geological history of Earth.

One impact responsible for worldwide destruction was caused by the projectile that did in the dinosaurs. The discovery of the dinosaurs, the realization that something wiped them out, and the search for the cause of their demise are the topics explored in the next chapter.

XII

DINOSAURS AND THE CRETACEOUS-TERTIARY EXTINCTION

O

> I returned, and saw under the sun,
> That the race is not to the swift,
> nor the battle to the strong,
> neither yet bread to the wise,
> nor yet riches to men of understanding,
> nor yet favor to men of skill;
> but time and chance
> happeneth to them all.
> —The Teacher, *Ecclesiastes 9:11*

FOSSILIZED dinosaur footprints were probably known to African and South American peoples in ancient times. In Brazil, local tribesmen examined the tracks, imagined the creatures that had made them, and carved images of giant running birds in nearby rocks. But it was not until 1677 that the first scientific illustration of a dinosaur bone was made; the illustrator was an Englishman named Robert Plot. The bone puzzled him, but he concluded that it was probably a bone of an elephant brought to England hundreds of years earlier, during the Roman occupation.

In the eighteenth and early nineteenth centuries, other dinosaur bones turned up in England and France; a few were also found in the western United States. In 1815 Reverend William Buckland of Oxford University dug up several large teeth and a fragment of a lower jaw; the fossils were very similar to (but much larger than) those of a monitor lizard. Buckland named the extinct creature *Megalosaurus* (Latin for great lizard). From the size of the teeth and jaw, Buckland's colleagues estimated that the animal was 12 meters long. As interesting as this animal was, it did not greatly exceed the length of modern reptiles — some crocodiles were known to be nearly 9 meters long.

Around 1820 Gideon A. Mantell, an English surgeon and gentleman paleontologist, found giant teeth that clearly resembled in shape (but dwarfed in size) those of a modern iguana. The extinct animal was called *Iguanodon*, and its discovery electrified the scientific community. A giant herbivorous reptile was indeed a new kind of animal.

By the mid-nineteenth century, new dinosaur fossils had been recovered in Germany, Switzerland, and Russia. Also unearthed were fossils of unrelated marine reptiles (ichthyosaurs and plesiosaurs) and flying reptiles (pterosaurs). By 1841 six different giant land-dwelling reptile species had been described, allowing zoologist Richard Owen to create a new group of reptiles, the Dinosauria. The animals of this group were characterized by their erect limbs and distinctly structured pelvises. It soon became clear that in Triassic, Jurassic, and Cretaceous times, huge reptiles had roamed the earth. The challenge was to determine what had happened to them.

The simple fact that dinosaurs are extinct is not surprising. In his 1993 book *Evolution of Life*, paleobiologist Michael L. McKinney estimated that since life originated on Earth three and a half billion years ago (some recent studies suggest that life started even earlier), there have been about three billion species. Because only five to fifty million species exist on Earth today, it is clear that more than 98 percent of all species that ever existed are now extinct.

Most extinct species have disappeared due to relatively mundane causes including predation, outcompetition, climate change, volcanic eruptions, the drying of a lake or river, or the burning of a forest. Extinctions of plants and animals arising from such causes occur in every geological period and are called background extinctions.

Superposed on the background extinctions are five mass extinctions—geologically brief periods of time when, all over the globe, large numbers of many kinds of species disappeared. These events occurred in the late Ordovician 438 million years ago, the late Devonian 380 million years ago, at the end of the Permian 251 million years ago, the late Triassic 205 million years ago, and at the end of the Cretaceous 65 million years ago.

The Cretaceous period was immediately followed by the Tertiary. Dinosaur bones occur in Cretaceous rocks but not in the overlying (hence younger) Tertiary rocks. The dinosaurs were not the only group to die out at the end of the Cretaceous. Among the many groups that vanished are the plesiosaurs, pterosaurs, ammonites (mollusks related to squids), many species and genera of snails, clams, and sea urchins, and some types of corals, bony fishes, and sponges. The tiny, single-celled, usually skeletal animals known as foraminifera come in two types—those that float on the ocean surface and those that dwell near the ocean bottom. Among the thirty-six floating foraminifera species, only one survived the Cretaceous-Tertiary (K-T) extinction; the bottom dwellers fared much better. All land vertebrates weighing more than 25 kilograms disappeared, but among the dinosaurs the extinction was total, wiping out even the relatively few smaller species. Many species of lizard also vanished. Perhaps 75 percent of the genera of marsupial mammals died out, including nearly all North American species.

Some species of turtles, salamanders, frogs, alligators, and crocodiles disappeared, but many survived. Also surviving were many species of birds, brachiopods (clamlike organisms sometimes called lampshells), gastropods (e.g., slugs, abalones), echinoderms (such as starfish, sea cucumbers, and crinoids), and crustaceans (e.g., crabs and shrimp).

Although more difficult to document, some plant species died off as well. There was a major redistribution of plant communities at the beginning of the Tertiary. Pollen from flowering plants declined from 75 percent to 1 percent of the total while fern spores increased fourfold, indicating that ferns opportunistically took over devastated landscapes.

The resolution of events that transpired 65 million years ago is apt to be poor. It may be difficult to tell if the end-Cretaceous extinctions occurred over a period of two million years, two thousand years, two years, or two months. Nevertheless, the answer is of immense importance in determining the cause of the extinction. If the cause was the impact of an asteroid, then different organisms should have died out at the same time, probably within a few months or years; if the cause was the tectonic separation or joining of continents, then extinctions should have occurred over millions of years, with some genera disappearing long before others.

The rareness of many types of fossils makes it difficult to resolve such time spans. If a mass extinction was abrupt and the fossils in a sequence of sedimentary rocks are plentiful, a paleontologist should find fossils of the doomed organisms having undiminished abundances in younger and younger rocks until the boundary is reached. At that point, the abundance of fossils should plummet to zero. But if the fossils are rare, it would be unlikely to find one right at the boundary; the youngest fossil is bound to be some distance below. In 1981 the uppermost (youngest) dinosaur bone that had been recovered was three meters below the beginning of the Tertiary; by 1997 the distance had been reduced to less than one meter.

The statistical distribution of rare, randomly distributed fossils in rocks was explored by paleobiologists Phillip Signor and Jere Lipps, then at the University of California at Davis. Their analysis, now termed the "Signor-Lipps effect," indicates that the probability of finding a particular fossil in rocks close to a boundary (between two distinct geological periods) decreases as the boundary is approached. This is because the amount of rock available to be searched decreases with increasing proximity to the boundary. Less rock means fewer fossils. An abrupt extinction could appear to be gradual because the fossils seem to peter out some distance below the boundary.

In 1986 Peter Sheehan and coworkers completed a rigorous study of the diversity of dinosaurs in eastern Montana and western North Dakota during the final 2.2 million years of the Cretaceous. They found that for most of this period, up until the last 250,000 years, the relative proportions of the different dinosaur families had remained the same. Because a gradual extinction is unlikely to have affected different dinosaur families equally, their constant proportions indicate that the extinction of all the dinosaurs was confined to the last 250,000 years of the Cretaceous. The data do not distinguish between an abrupt extinction on the last weekend of the Cretaceous or a relatively gradual one that took every day of the final 250,000 years.

The 1940 Disney movie *Fantasia* depicted the manner in which contemporary scientists thought the dinosaurs had died. Climate change did them in: Earth got hotter, swamps gave way to deserts, and previously reliable water holes dried up. Some later workers offered alternative suggestions: perhaps mammals ate dinosaur eggs (presumably poached); maybe dinosaurs suffered from testicular heating or were simply unfortunate victims of a "failed experiment." Perhaps they grew too large to mount their mates or developed a fondness for eating poisonous flowers. The dinosaurs' demise has also been attributed to slipped disks in their vertebral columns, blindness caused by cataracts, viral epidemics, parasites, arthritis, bone fractures, excessive stupidity, constipation, and overpredation. It has even been suggested that the entire dinosaur lineage grew less able to compete with mammals and died off.

But by focusing on the disappearance of the dinosaurs, researchers were asking the wrong question. Stupidity, egg stealing, constipation, and arthritis would not cause foraminifera, ammonites, marsupial mammals, and sponges to die off at the same time as dinosaurs. Because about 65–70 percent of all living species became extinct at the end of the Cretaceous, global catastrophes must be considered. The study by Sheehan and coworkers is consistent with an abrupt extinction event, as is the distribution of fossils in the late Cretaceous as interpreted in light of the Signor-Lipps effect. Sudden global cataclysms include the impact of a large asteroid, a supernova explosion, a dramatic spurt of volcanic activity, and simultaneous worldwide forest fires.

Unsupported speculation about the cause of the K-T mass extinction began to wind down in 1980 with a publication in *Science* by Nobel laureate physicist Luis Alvarez, his son, geophysicist Walter Alvarez, and their colleagues at the University of California, Berkeley. They reported the discovery of iridium-enriched clay at several K-T boundary sites. Dutch paleontologist Jan Smit had previously had some K-T boundary samples analyzed but was too sick with mononucleosis to examine the data. While recuperating, he heard about the Alvarez work and found that his samples were also enriched in iridium.

The K-T boundary itself consists at many sites of a centimeter-thick, fossil-free clay layer wedged between the uppermost Cretaceous limestone rock (containing large Cretaceous-age foraminifera) and the lowermost Tertiary limestone (containing much smaller, Tertiary-age foraminifera). The Alvarez team found that the iridium concentration in this clay layer is about 9 parts per billion. Although this is an extremely small value, it is thirty times larger than the average value of 0.5 parts per billion in crustal rocks. (The very low concentration of iridium in the crust is a result of core formation early in Earth's history, when the planet largely melted and heavy metals like iridium joined metallic iron and sank to Earth's center.)

Within a few months of the Alvarez paper, three other research teams reported iridium-enriched clay at other marine (i.e., ocean-deposited) sites, and within a year Carl Orth of Los Alamos National Laboratory found an iridium anomaly in nonmarine rocks in New Mexico. Within two decades, more than one hundred K-T sites around the world were found to have iridium enrichments.

Three different mechanisms were proposed as the cause of the high concentrations of iridium at the K-T boundary. (1) Some types of meteorites contain as much as 500 parts per billion iridium; the impact of a multikilometer asteroid could have distributed iridium-rich dust of the observed concentration all over the globe. (2) Iridium and all other elements heavier than iron (i.e., with atomic numbers greater than 26) were originally created in supernova explosions. It therefore seemed plausible that a nearby supernova could have deposited the iridium and unleashed fatal radiation, causing a mass extinction. (3) A massive volcanic eruption fed by a deep-seated magma chamber that scavenged iridium from the lower mantle might have injected enough dust into the atmosphere to block sunlight, cool the climate, cause the mass extinction, and distribute iridium around the earth. The 1983 discovery of iridium-rich airborne particles in volcanic emissions from Kilauea in Hawaii is consistent with the volcanic hypothesis.

Alvarez and coworkers were able to eliminate the supernova hypothesis almost immediately. Such an explosion should have showered Earth with plutonium 244, a radioactive isotope with 94 protons and 150 neutrons. Because the half-life of this isotope is 83 million years, plenty of it should still be around after 65 million years. Analysis of the clay revealed no plutonium 244, contradicting a key prediction of the supernova hypothesis.

Russian physicist George Bekov and Frank Asaro of Lawrence Berkeley Laboratory measured the concentrations of iridium and two chemically similar metals (ruthenium and rhodium) in the K-T boundary clay and found that the relative proportions of these elements were approximately the same as in chondritic meteorites and very different from those in volcanic emissions. These data made the volcanic hypothesis less tenable and supported the asteroid-impact scenario.

Additional evidence for an asteroid collision began pouring in.

At K-T boundary sites throughout North America, grains of shocked quartz were recovered. These grains have distinct sets of linear structures formed by a pulse of very high pressure. Shocked grains of the minerals feldspar (an alkali- and aluminum-rich silicate) and zircon (a zirconium-rich silicate) have also been discovered. K-T boundary sites in Haiti contain abundant, altered glass spherules that probably formed from rapidly cooled airborne impact-melt droplets. The ratios of two different isotopes of osmium (osmium 187 to 188) in K-T boundary clays range from about 0.14 to 0.21, similar to the ratios in meteorites and much lower than typical values in crustal rocks (0.7–1.0). Wendy Wolbach of the University of Chicago and her colleagues reported charcoal and soot associated with the iridium-enriched clay at many K-T boundary sites. These carbon compounds were interpreted as resulting from global wildfires.

Meixun Zhao and Jeffrey Bada of the Scripps Institution of Oceanography reported extraterrestrial amino acids in K-T boundary sediments; the amino acids they detected (α-amino-isobutyric acid and equal amounts of left- and right-handed isovaline) are extremely rare on Earth but abundant in carbonaceous chondrites. Luann Becker, then at the University of Hawaii, and her colleagues found soccer-ball-shaped carbon molecules called fullerenes in the Allende and Murchison carbonaceous chondrites, in rocks from the Sudbury, Canada, impact structure, and in the K-T boundary clay layer. The fullerenes act like cages, trapping molecules of noble gases; the ratios of helium 3 to helium 4 in these fullerenes are about ten times higher than the maximum terrestrial mantle values and thus indicate the presence of an extraterrestrial carbon component at the K-T boundary.

In the years following 1980 most astronomers, geologists, meteorite researchers, and invertebrate paleontologists came to accept the hypothesis that the end-Cretaceous mass extinction was caused by the impact of a large asteroid or comet. But the impact of a 10-kilometer projectile would be expected to produce a crater 100–200 kilometers in diameter. Where was the crater? Some researchers suggested that the crater had formed on the ocean floor but, because of movements of Earth's tectonic plates, had been drawn down into the mantle during the last 65 million years. Other researchers pressed on. Several candidate craters were suggested including the Kara Crater in Russia and the Manson Structure in Iowa, but these proved to be too old (72 and 74 million years old, respectively).

If the impact had been in the ocean, submarine landslides would have carried seafloor sediments, rock fragments, and impact glass away from the impact site. A huge wave called a tsunami would have been generated, tearing up the seafloor and carrying the eroded sediment along with it. As the tsunami approached land, it would have built itself up into a kilometer-high wave that blasted into the coastal landscape. The resulting deposit would contain coastal sand, jumbled rock fragments, glass spherules, and torn-up seafloor sediments. Such a locality at the K-T boundary, complete with an iridium anomaly, was identified at the Brazos River on the southern peninsula of Haiti. Sixty-five million years ago, Haiti's position in the Caribbean Sea was not that different from its present location. Canadian geologist Alan Hildebrand, then at the University of Arizona, reasoned that because the Gulf of Mexico and Caribbean Sea were largely closed bodies of water, the tsunami must have been generated somewhere between Texas and Colombia. The K-T crater must be there too.

After painstaking detective work, Hildebrand came across the scant information available about gravity anomalies associated with a buried circular structure on the Yucatán Peninsula in Mexico. This structure had been identified as an impact crater by Antonio Camargo and Glen Penfield, geologists working for PEMEX, the Mexican national oil company. Their work was published as an abstract of a meeting of the Society of Exploration Geophysicists, but K-T boundary researchers were unaware of the report. It was not until Hildebrand met Camargo and Penfield and published a paper in *Geology* in 1991 that the world became aware of the Chicxulub (pronounced "cheek-shew-loob")

Crater. Planetary scientist David Kring points out that Chicxulub is a Mayan word roughly translated as "tail of the devil." The crater is between 180 and 300 kilometers in diameter, making it the largest impact crater known on Earth. Half of the crater occurs offshore in the Gulf of Mexico. Impact-melt rock from the crater was dated at 65.0 million years. The Chicxulub Crater was thus in the right place (able to account for the tsunami deposits in Haiti), of the right size (implying that there would have been sufficient impact energy to cause widespread havoc and deposit iridium-rich dust around the globe), and of the right age to be responsible for the K-T mass extinction.

The Yucatán Peninsula is composed of continental crust overlain by sediments containing calcium carbonate (calcite; $CaCO_3$), calcium-magnesium carbonate (dolomite; $CaMg(CO_3)_2$) and calcium sulfate (anhydrite; $CaSO_4$). Sixty-five million years ago the site was covered by shallow ocean water.

The impact of the 10-kilometer projectile into the Yucatán had severe and diverse environmental consequences, both immediate and long-term. The tremendous energy liberated by the impact (equivalent to about 100 million megatons of TNT) melted the local continental crust and overlying sediments, excavated material from as deep as 14 kilometers, injected enormous clouds of steam into the air, and ejected billions of tons of rock fragments above the atmosphere. Airborne droplets melted by the impact cooled into glass spherules before falling to the ground. Seismic waves raced across Earth's surface (and through Earth's interior), causing violent shaking and submarine landslides. Large tsunamis were generated, spreading across the Gulf of Mexico and obliterating coastal forests. Huge boulders, rock fragments, and sand were deposited ashore. The air blast scoured vegetation off the peninsula and pushed out over the Atlantic and Gulf of Mexico.

Ballistic ejecta that had been launched into near space reentered the atmosphere and were heated by friction with air molecules. The heated air became incandescent, transmitting infrared radiation to Earth's surface, broiling plants and exposed animals. The high temperatures ignited trees and brush, creating continent-size wildfires. The air grew hot enough to break the bonds of molecular nitrogen (N_2) in the atmosphere to form nitric oxide (NO). The NO molecules reacted with oxygen and water vapor to form droplets of nitric acid (HNO_3). Calcium sulfate from the Yucatán sediments was blasted into the atmosphere as sulfate aerosols. These particles reacted with water vapor to form sulfuric acid (H_2SO_4). The nitric and sulfuric acid droplets fell as acid rain, killing animals and plants and dissolving carbonate rocks. Also dissolved were the carbonate shells of marine organisms such as foraminifera that dwelled in the uppermost 100 meters of the oceans. Burning forests and dissolved carbonates released huge amounts of carbon dioxide (CO_2) into the atmosphere.

Billions of tons of dust were injected by the impact into the stratosphere, plunging the world into darkness and severely hindering photosynthesis in land plants and algal plankton at the ocean surface. Temperatures plummeted to subfreezing levels and remained low for months. As the dust settled and the darkness ebbed, temperatures slowly increased. The abundant carbon dioxide in the atmosphere acted as a greenhouse gas, trapping infrared radiation and raising global temperatures to sweltering levels. Because carbon dioxide is removed from the atmosphere only slowly, temperatures probably remained high for thousands of years.

Climatologist Peter Fawcett of the University of New Mexico and physicist Mark Boslough of Sandia National Laboratories suggested that billions of tons of debris were lofted into Earth orbit. Within one hundred thousand years of the impact, long after surface temperatures had returned to normal, the orbiting debris may have formed itself into a wide ring resembling that of Saturn. The ring would have cast deep shadows across Earth's surface, causing substantial cooling. The ring may have lasted two or three million years until the orbiting dust fell back to Earth and temperatures returned to normal once again.

It is easy to see that such catastrophic events would cause a mass extinction. Animals and plants would be killed, land dwellers as well as ocean dwellers, organisms that inhabited the tropics as well as those that lived in polar regions. Species that persevered were not necessarily tougher than those that perished; they just happened to occupy environmental niches that allowed sufficient numbers of individuals to survive. Animals that dwelled in caves or spent a lot of time underground had an advantage over those that crawled through brush. Animals that lived on the ocean floor were less vulnerable than plankton floating at the surface. Animals that stored their food fared better than those that did not.

As more geologists accepted an impact as the cause of the K-T extinction, many began to wonder if the four other mass extinctions were also attributable to collisions. Although there were no confirmed reports of iridium anomalies at the other boundaries, some scientists grasped at the impact hypothesis because no other mechanism had been shown to be capable of causing a mass extinction. In 2001 geochemist Luann Becker and her colleagues analyzed sediments from the Permian-Triassic boundary and reported fullerenes containing trapped helium and argon with isotopic ratios similar to those of carbonaceous chondrites but different from those of Earth. These results are consistent with an impact trigger for the Permo-Triassic extinction event, but as of this writing, unequivocal confirmatory evidence of a large impact is lacking.

In 1984 David Raup and Jack Sepkoski from the University of Chicago plotted significant extinction events over time and perceived a 26-million-year periodicity in the data. Astronomers Marc Davis and Piet Hut and physicist Richard Muller suggested that the Sun might have a companion star (which they nicknamed Nemesis) in an elliptical orbit that comes close enough to the Oort Cloud of comets every 26 million years to perturb some of them toward the inner solar system. Periodic comet showers would result. Walter Alvarez and Richard Muller reasoned that dated impact craters should show the same periodicity; they conducted a literature survey and concluded that such a periodicity was probable.

However, most impact craters are dated rather imprecisely, and the majority of researchers saw no periodicity in the crater ages. Nemesis has still not been found, and computer simulations indicate that its orbit would be unstable over the age of the solar system.

An implicit prediction of the comet-shower hypothesis is that there should also be a 26-million-year periodicity in iridium spikes in continuously deposited sediments. UCLA geochemists Frank Kyte and John Wasson measured the iridium concentrations of 149 samples from a deep-sea piston core containing clay that was deposited continuously between 67 and 33 million years ago. The profile of the iridium concentrations shows a huge iridium spike in the clay deposited 65 million years ago but no spikes anywhere else (figure 12.1). In particular, clays deposited 39 million years ago and 13 million years ago show no iridium spike, contrary to the periodic-comet-shower hypothesis.

Most researchers eventually concluded that mass extinctions are not periodic. Nevertheless, the question remained as to whether the K-T projectile was a lone comet or a lone asteroid.

In 1998 Scripps researchers Alexander Shukolyukov and Gunter Lugmair measured the chromium isotopic compositions of various terrestrial rocks and meteorites and found appreciable differences among them. They also measured K-T boundary samples and found a chromium isotopic composition consistent with the projectile having been a carbonaceous chondrite.

Three weeks after this paper was published, Frank Kyte reported finding a 2.5-millimeter fossil meteorite (figure 12.2) in a North Pacific deep-sea drill core that contains sediments from the K-T boundary. Evidence that this small rock is a piece of the long-sought K-T projectile is strong. Although tiny, the rock is one thousand times larger than anything else in the sediment. Kyte showed that 65 million years ago, the rock's recovery location was thousands of kilometers from the nearest continent. The rock is therefore unlikely to have been deposited by the wind; it must have fallen as a meteorite. Analysis of the sample showed that it differs significantly in composition from the surrounding

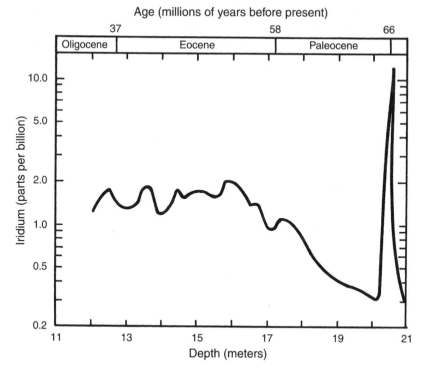

Figure 12.1. The concentration of iridium (Ir) in ocean-floor clay deposited between 33 and 67 million years ago. Apparent differences in iridium at depths between 12 and 20 meters reflect variations in sedimentation rate. The only spike indicating very high concentrations of iridium occurs at the Cretaceous-Tertiary boundary. (Diagram after Kyte and Wasson (1986), *Science* 232, 1225–29)

K-T boundary sediments: it is highly enriched in the metals iridium, gold, and chromium. The iridium concentration in the rock is characteristic of chondritic meteorites, and the iron and chromium concentrations are within a factor of two of those in chondrites. Although the original minerals in the fossil meteorite have been replaced by iron oxide and clays, the sample has retained portions of its original chondritic texture. The rock resembles a metallic iron-nickel-rich carbonaceous chondrite. This conclusion is consistent with the occurrence of helium 3–rich full-

Figure 12.2. A highly weathered carbonaceous chondrite meteorite (broken into two pieces) and surrounding clays found in sediments collected at the Cretaceous-Tertiary boundary on the floor of the North Pacific Ocean. Maximum dimension of the meteorite is 2.5 mm. (Courtesy of Frank Kyte, UCLA)

erenes and extraterrestrial amino acids in K-T boundary clays, and Shukolyukov and Lugmair's interpretation of the chromium-isotopic data.

Because many comets are in highly eccentric, high-inclination orbits, most of those that strike Earth collide at high velocities, resulting in complete vaporization of the projectile. The fact that the fossil meteorite fragment exists at all is evidence that it is from an asteroid (or at least not from a comet in a typical orbit).

A few biologists and vertebrate paleontologists still doubt that the impact of the K-T projectile caused the mass extinction. They argue instead that the effects of the impact, while severe, were largely local. They prefer strictly terrestrial extinction mechanisms such as a precipitous decline in sea level, a change in the salinity of the ocean, or a massive volcanic eruption. It is cer-

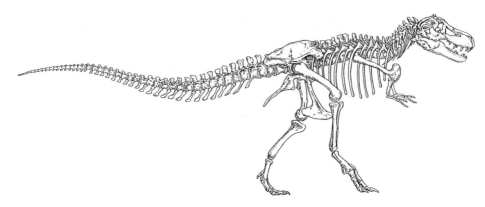

Figure 12.3. What all the fuss was about—skeletal remains of extinct fearsome creatures such as this *Tyrannosaurus rex*. (Drawing by Dorothy Norton, courtesy of Black Hills Institute of Geological Research)

tainly possible that such mechanisms triggered other mass extinctions, but they are not the ultimate cause of the K-T extinction. What killed the dinosaurs (figure 12.3) was the impact of a large rock from the asteroid belt. As Frank Kyte said in a 1999 interview, "It was one of the worst days the Earth has had in the last billion years." Nothing has been the same since.

Since the collision of the projectile that ended the Cretaceous 65 million years ago, there have been numerous less-energetic impacts. Some of the more notable recent impacts in the solar system are explored in the next chapter.

XIII

RECENT IMPACTS:
TUNGUSKA TO
SHOEMAKER-LEVY 9

$$\bigcirc$$

On this day, the sun
Appeared—no, not slowly over the horizon—
But right in the city square.
A blast of dazzle poured over,
Not from the middle of the sky,
But from the earth torn raggedly open.
   —Agyeya, *Hiroshima*

COLLISIONS among solar-system objects are not phenomena restricted to the ancient past. Small particles impact
every planet, every asteroid, every comet, and every moon every
day. A comet runs into the Sun every few years, and sofa-size
meteoroids plunge into Earth's atmosphere once a month.

There are about five hundred to one thousand asteroids at least
1 kilometer in size in Earth-crossing orbits; there are a few hundred thousand 100-meter-size asteroids and several million 10-
meter-size asteroids in Earth-crossing orbits. Given the large pop-

ulation of near-Earth asteroids, it is not surprising that there have been some recent near misses. In December 1994 a 10-meter asteroid, 1994 XM1, came within 112,000 kilometers of Earth. In May 1996 a 200-meter asteroid, 1996 JA1, came within 461,000 kilometers. On 7 August 2027 a much larger asteroid, 1.1-kilometer-diameter 1999 AN10, will pass within 389,000 kilometers of Earth, just beyond the mean distance of the Moon (384,000 kilometers).

The closest near miss in modern times came in the early afternoon of 10 August 1972. A 13- to 80-meter near-Earth asteroid traveling at 15 kilometers per second traversed 1,500 kilometers of Earth's atmosphere at a very low, nearly horizontal angle. The body was first observed over southwestern Utah heading north through Idaho, Montana, and then Alberta, Canada. It was seen by thousands of startled eyewitnesses (including my future wife, who was driving across Montana with her family). The asteroid came to within 58 kilometers of the surface over Montana, where sonic booms were heard; it skipped back out of Earth's atmosphere over Alberta at an altitude of 102 kilometers. It left a trail of smoke behind it that was visible for 30 minutes.

If the angle of entry had been somewhat steeper, it might have detonated in the atmosphere with an energy yield of a few megatons of TNT, creating an explosion several hundred times more energetic than the atomic bomb dropped on Hiroshima. (I might have ended up single.)

But sometimes near-Earth objects don't miss.

At 7:17 A.M. local time on 30 June 1908 an explosion occurred over the subarctic evergreen forest of central Siberia near the Stony Tunguska River. About 15 megatons of energy were released 9–10 kilometers above the surface. Witnesses reported a huge, "blindingly bright," blue-white fireball with a glowing tail and thick dust trail. They heard loud crashes, crackling, and rumbling and described the sounds as thunderclaps, artillery, and distant gunshots. A fierce hot wind killed grazing reindeer, lifted up reindeer herders, charred their clothes and dashed the herders to the ground. Tents and storage huts were blown down, and sur-

Figure 13.1. Felled trees in the Tunguska forest caused by the explosion of a friable projectile over Siberia on 30 June 1908. This site is about 8 km from the center of the devastation. Trees were blown down in a radial pattern from ground zero. (Photograph by L. A. Kulik, courtesy of the Smithsonian Institution)

face vegetation was scorched. Sixty kilometers away, windows were shattered, dishes were broken, and household utensils were knocked off shelves. Trees immediately beneath the blast remained upright but were singed and stripped of their branches and leaves. Starting at 5 kilometers from ground zero, trees were felled throughout a region of about 250 square kilometers; the tops of the tree trunks pointed directly away from the explosion site (figure 13.1). Widespread destruction occurred over an area of 2,200 square kilometers.

Sound was heard and seismic waves were recorded more than 1,000 kilometers away. Air pressure waves spread around the globe twice. For several nights thereafter, observers in Europe and western Asia noted a red glow in the upper atmosphere and puzzled over the "light nights" of unending twilight. Bright silvery high-altitude clouds were seen. Over the next two months the light nights gradually grew darker, and nighttime newspaper readers had to light their lamps.

It was nineteen years before the first scientific expedition reached the site. Although a meteorite impact was suspected, no crater was located and no meteorites were found. Leonid A. Kulik, leader of the 1927 expedition, interviewed witnesses, discovered the flattened forest, and mapped the devastated landscape. He reported that "the results of even a cursory examination exceeded all the tales of the eye-witnesses and my wildest expectations." Kulik led three more expeditions to the site, supervised aerial photography of the region, drilled holes in the frozen ground, and used a magnetometer in unsuccessful searches for meteorite fragments.

Over the years, fantastic theories were concocted to explain the event: mini–black holes, nuclear explosions, crashing alien spacecraft, and antimatter meteoroids. Less imaginative researchers suggested a friable, underdense (0.001–0.01 gram per cubic centimeter) comet had burst in the atmosphere. Others proposed that the object was a common stony meteorite that had exploded.

The most comprehensive study of the nature of the projectile was made in 1993 by planetary scientist Christopher Chyba and colleagues. They examined the effects of aerodynamic forces on objects a few tens of meters in radius that entered Earth's atmosphere at hypersonic velocities at different angles. They found that the fate of projectiles in this size range is governed mainly by their composition:

1. Comets explode at high altitudes where the aerodynamic strain exceeds the comets' material strength.

2. Carbonaceous chondrites, less friable than comets but more friable than stony meteoroids, explode at somewhat lower altitudes.

3. Metallic iron projectiles less than 30 meters across scarcely fragment in the atmosphere at all; they tend to remain intact until reaching very low altitudes. When disruption finally begins, their fragments remain close together and impact the surface like a single coherent projectile.

4. Common, noncarbonaceous stony meteorites (such as ordinary chondrites or meteoritic basalts) entering Earth's atmosphere at angles of 30° or more can be fractured by aerodynamic forces and spread apart. The new small individual fragments would experience greatly increased atmospheric drag. This would cause further separation of the fragments, more rapid deceleration, and catastrophic detonation at an altitude of about 9 kilometers. This fourth scenario is consistent with the fate of the Tunguska projectile.

Chyba and colleagues calculated that the rising plume of heated air from the Tunguska airburst would have reached altitudes of 40–50 kilometers and entrained large amounts of water. Dispersal of the water by winds in the upper atmosphere could have produced the silvery high-altitude water-ice clouds responsible for the light nights after the explosion.

Three post-Tunguska terrestrial impacts serve to illustrate the idea that for projectiles smaller than a few tens of meters in size, composition is the main variable governing the altitude of projectile disruption. (Larger projectiles tend not to disrupt at all until they reach Earth's surface.)

Large metallic iron projectiles disrupt near the surface. At 10:38 A.M. local time on 12 February 1947 a large iron meteorite fell in Siberia near the Sikhote-Alin Mountains (figure 13.2). Several kilotons of energy were released. A bright fireball with a smoky trail was observed traveling north to south. Sonic booms were heard. In villages near the fall, windowpanes fell out, doors flew open, and smoldering cinders jumped out of burning stoves. More than 120 meteorite craters, ranging in diameter from 0.5 to 26.5 meters, were located within a dispersion ellipse of 4 × 12 kilometers. This small dispersion indicates that the projectile fragmented at a very low altitude, estimated to be about 4.5 kilometers. The largest intact specimen had a mass of 1,745 kilograms and, altogether, more than 28,000 kilograms of iron meteorite were recovered. The estimated total mass of material that reached the ground (including meteoritic dust mixed in the soil) is about 100 metric tons.

Figure 13.2. A 106-kg piece of the Sikhote-Alin iron meteorite that fell in mountain forests north of Vladivostok, Russia, on 12 February 1947. (Courtesy of Walter Zeitschel and *Meteorite* magazine)

Large carbonaceous chondrites explode at moderately high altitudes. At 9:47 P.M. local time on 31 March 1965 a projectile traveling at an angle of about 15° exploded 30 kilometers above British Columbia, Canada, near the town of Revelstoke. Estimates of the energy of the explosion range from 20 to 70 kilotons. Only about 1 gram of meteoritic material was recovered from the snow atop a frozen lake. Analysis showed that it was a CI carbonaceous chondrite—the most friable meteorite type and one that contains about 17 percent water bound in silicate minerals. The meteorite was named Revelstoke, after the nearby town.

Cometary projectiles explode high in the atmosphere. The Kincardine fireball, observed on 17 September 1966 at 8:47 P.M. local time, exploded over Ontario, Canada, at an altitude of just over 60 kilometers. The energy of the explosion may have been as much as a megaton. No material was ever recovered, and it seems likely that the projectile was a small comet.

Although Earth is bombarded by numerous objects tens of meters in size, each of which explodes with energy in the multi-kiloton range, the disruption of many of these projectiles high in the atmosphere protects us from frequent disasters. Smaller non-friable objects tend not to be catastrophically disrupted and do impact more frequently, but generally pose no threat. Larger objects pass through the atmosphere unscathed and pose a significant danger; fortunately, however, they are rare.

The most energetic impact events ever witnessed by humans occurred over a six-day period in July 1994. The target was Jupiter, and the projectile was an unusual comet known as Shoemaker-Levy 9.

The comet was discovered by Carolyn S. Shoemaker at Palomar Observatory on 25 March 1993 as she looked under a stereomicroscope at negatives of the star field near Jupiter. The photographs had been taken with an 18-inch (0.46-meter) wide-field photographic telescope by her coworker, amateur astronomer and science writer David H. Levy. Shoemaker saw "a bar of light with coma and several tails" in the image that was about 50 arc-seconds (about 0.14°) across; she declared that it looked like a "squashed comet." Gene Shoemaker (Carolyn's husband) and David Levy also examined the image and reported the misshapen comet to Brian Marsden. Marsden is the director of the International Astronomical Union's Central Bureau for Astronomical Telegrams at the Smithsonian Astrophysical Observatory in Cambridge, Massachusetts. His office is the clearinghouse for astronomical discoveries.

After being alerted to the discovery, astronomer Jim Scotti photographed the new comet later that night with the 36-inch (0.91-meter) Spacewatch camera at Kitt Peak National Observatory in Arizona. He noted "at least five discrete comet nuclei side by side." Marsden named the new comet Shoemaker-Levy (1993e). It eventually would be called Comet P/Shoemaker-Levy 9, eight previous periodic comets (i.e., comets with orbital periods less than 200 years) having been discovered by the Shoemaker-Levy team.

As excitement intensified among astronomers, hundreds of additional images were taken. Astronomers Jane Luu and David Jewitt used the University of Hawaii's 88-inch (2.2-meter) reflector and resolved seventeen discrete nuclei. Jewitt described them as "pearls on a string," a metaphor that proved popular with the media. A handful of prediscovery photographs were located, and the orbit of the comet was determined. Refinements in the orbit allowed astronomers to piece together the recent history and imminent future of the comet.

Shoemaker-Levy 9 (SL9) was found to be a member of the Jupiter family of comets. About fifty such comets are known. These are typically low-inclination objects that orbit the Sun with periods of six to eight years. Their orbits are tied to that of Jupiter. In some cases, their aphelia (the farthest points in their orbits) are near Jupiter's orbit; in other cases, the point where the orbit intersects the ecliptic is near Jupiter's orbit. The comets were transformed from somewhat-longer-period objects (probably derived from the Edgeworth-Kuiper Belt between 34 and 45 AU from the Sun) to shorter-period ones by passing close to Jupiter. The giant planet's enormous gravity pulled the comets into these smaller orbits. Some Jupiter-family comets assume orbits around Jupiter for a few years or a few decades before being flung out into new orbits around the Sun. At any given time, there is probably one kilometer-size comet in temporary orbit around Jupiter.

Extrapolating SL9's orbit back into the past, Jet Propulsion Laboratory (JPL) scientists Paul Chodas and Donald Yeomans concluded that the comet was probably captured by Jupiter in 1929. After orbiting Jupiter for sixty-three years, SL9 passed within 21,000 kilometers of the cloud tops on 7 July 1992, far within the planet's Roche limit (the distance where a low-strength fluid body would disrupt, in Jupiter's case, about 103,000 kilometers from the cloud tops). Tidal forces, pulling slightly more on the side of the comet facing Jupiter than on the opposite side, ripped the friable nucleus apart. Eventually, SL9 separated into at

Figure 13.3. Comet Shoemaker-Levy 9 in early 1994 showing about twenty-one fragments and associated halos of reflected dust. (Image from the Hubble Space Telescope courtesy of the Space Telescope Science Institute and Paul M. Schenk, Lunar and Planetary Institute)

least twenty-one fragments and accompanying swarms of dust (figure 13.3). The fragments were labeled alphabetically. This was to be the comet's final orbit; its closest approach to Jupiter on the next pass in July 1994 would see it collide with the planet's Southern Hemisphere.

Plans were made to observe the impacts with terrestrial telescopes in visible light and at infrared and radio wavelengths. The Kuiper Airborne Observatory (KAO), a large C-141 aircraft, would climb to 41,000 feet (12.5 kilometers) and train its telescopes on Jupiter. The Hubble Space Telescope, the International Ultraviolet Explorer, and the ROSAT satellite/x-ray telescope would observe the impacts from Earth orbit. Although it ultimately failed to see any changes on Jupiter, the distant *Voyager 2* spacecraft (at that time 40 AU from Earth) would crank up its radio experiments and also give it a try. Finally, the *Galileo* spacecraft, already en route to Jupiter (but still seventeen months away), would observe the impacts more directly.

The only problem was that the impacts would occur just beyond Jupiter's limb as seen from Earth. We would not see the actual impacts. If plumes were to rise above the impact sites, those might be visible; if any scars were to remain in the Jovian atmosphere twelve to twenty-four minutes after the impacts, they might be seen as they rotated into view.

Prior to the first impact of fragment A on 16 July 1994, there was great uncertainty about whether astronomers would see anything at all. At one extreme were predictions that flashes from the impacts would be so bright that Jupiter would be visible in daylight. Some suggested that impact flashes would be reflected off Jupiter's satellites, causing them to brighten temporarily. Among the extremists on the other side was JPL comet expert Paul R. Weissman. His article "Comet Shoemaker-Levy 9: The Big Fizzle Is Coming" appeared in *Nature* two days before the first impact. After the impacts had proven to be spectacular, Weissman explained that "fizzle" is really a Yiddish word meaning "great big humongous Jupiter-shaking comet explosion."

As the months passed, the fragments separated further, reaching a maximum span of about 900 arcseconds (0.25°) prior to impact. Some individual fragments (e.g., J and M) disaggregated into diffuse clouds of dust and disappeared. A few fragments (e.g., G, P, and Q) broke up into additional pieces.

Three groups of researchers awaited the results of the impact. Comet experts thought they could learn about the structure of comets by watching SL9 break apart as it neared Jupiter. They were eager to see the development of comas and tails on new subfragments. Jupiter aficionados thought they could learn something about the composition of the planet's atmospheric layers when material was ejected in the impact plumes. Some also hoped that the impacts would generate a seismic wave that would reflect off the planet's mantle of liquid metallic hydrogen and provide information on its size. Impact physicists were excited to witness a planetary collision and test their computer models against reality.

The larger comet fragments would penetrate to lower depths than the smaller ones. However, the sizes of the individual fragments were unknown; estimates ranged from a few hundred meters to a few kilometers. After the impacts, most researchers favored the smaller size ranges, but great uncertainty remained.

For six days, from the impact of fragment A on 16 July to the impact of fragment W on 22 July, astronomers, both professional and amateur, observed the collisions with great excitement. Overly optimistic astronauts aboard the space shuttle *Columbia* watched Jupiter with binoculars from Earth orbit.

The comet fragments slammed into Jupiter's atmosphere at speeds of 60 kilometers per second. Their rapid deceleration caused them to flatten, fluidize, and fragment. The fragments exploded, each releasing energy of the order of 10,000 megatons. Larger fragments such as G released appreciably more energy, perhaps approaching 100,000 megatons. The impacts generated hot, compressed gaseous plumes that expanded violently upward at 10 kilometers per second through the comet's wake in the atmosphere (figure 13.4). This was the tunnel created by the comet's downward passage that was temporarily filled with low-density gas. Plumes from the larger impacts reached heights of more than 3,000 kilometers above the cloud tops. Huge dark partial rings (larger than Earth) developed around the impact sites as fine-grained debris dropped out of the plumes onto Jupiter's ammonia clouds. The reentering material heated the atmosphere to incandescence, producing prodigious amounts of infrared radiation.

The impact scars persisted as Jupiter continued to rotate every 10 hours. Some scars lasted two weeks; others were readily observable for more than a month. (Some impact debris particles were still detectable in the turbulent Jovian atmosphere two years later.) As more comet fragments plunged into Jupiter's atmosphere, some impacts occurred adjacent to or on top of previous impact scars (figure 13.5). This was a new phenomenon that David Levy dubbed "impact clutter." It added to the difficulty of discerning the long-term atmospheric effects of individual impacts.

As the comet fragments passed through Jupiter's magnetosphere, huge bursts of ultraviolet energy were released. Intense auroral activity was detected in Jupiter's Northern Hemisphere, opposite the impact sites. The activity was due to charged particles becoming entrained in Jupiter's magnetic lines of force and funneled to the opposite hemisphere. Earth-based radio tele-

Figure 13.4. The plume above Jupiter's clouds caused by the impact of fragment G into Jupiter on 18 July 1994. (Image from the Hubble Space Telescope courtesy of the Space Telescope Science Institute)

scopes revealed that the impacts of fragments K and L induced temporary changes in Jupiter's radiation belts.

Data on the composition of Jupiter's atmosphere started pouring in. The Hubble Space Telescope detected sulfur, and the scientists deduced that it was derived from a hypothetical layer of ammonium hydrosulfide beneath the uppermost ammonia cloud layer. Carbon disulfide was also detected; this is a compound formed when hydrogen sulfide reacts with methane at high tem-

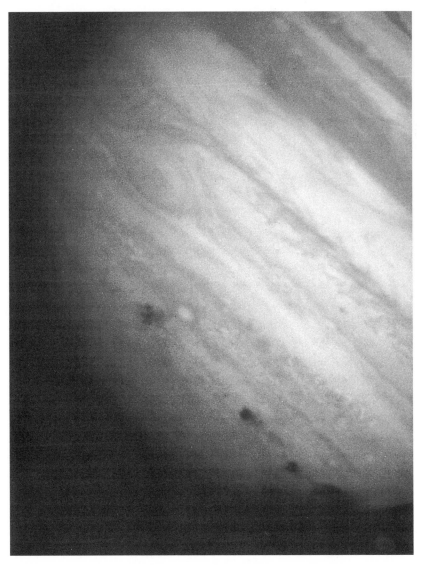

Figure 13.5. Multiple impact scars in Jupiter's atmosphere photographed on 22 July 1994. (Image from the Hubble Space Telescope courtesy of the Space Telescope Science Institute)

peratures and pressures like those generated in an impact. Water vapor was detected by the Kuiper Airborne Observatory twelve minutes after the impact of fragment R, but it was not clear if the water had come from Jupiter's atmosphere or the comet fragment.

Astronomer Thomas Hockey dug through historical records and found a handful of descriptions and drawings of large spots on Jupiter dating back to 1690. Some of the spots were reported to have changed their appearance over the course of a few days. Although intrinsic atmospheric disturbances cannot be ruled out, at least some of the spots may have resulted from unobserved SL9-like impacts.

At least occasionally during the history of the solar system, tidally disrupted comets in Jupiter orbit ought to have run into a

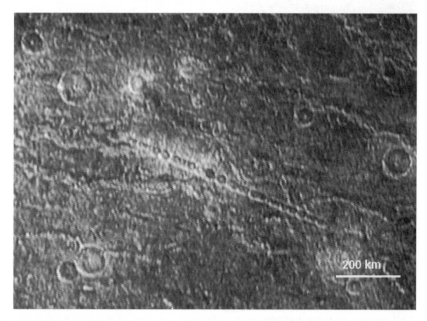

Figure 13.6. A 350-km-long crater chain known as Gomul Catena on Jupiter's moon Callisto photographed by the *Voyager 1* spacecraft. The chain was probably formed by multiple impacts of a tidally disrupted comet. (Courtesy of Paul M. Schenk, Lunar and Planetary Institute)

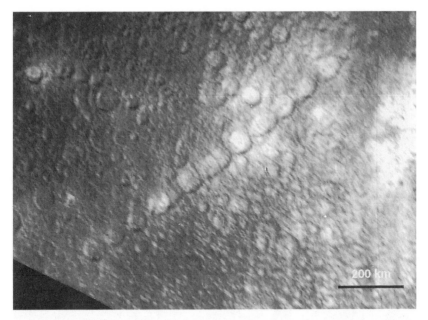

Figure 13.7. A 620-km-long crater chain known as Gipul Catena on Callisto taken by the *Voyager 1* spacecraft. (Courtesy of Paul M. Schenk, Lunar and Planetary Institute)

moon instead of Jupiter. Planetary scientist Jay Melosh proposed that crater chains called catenae on Jupiter's moons Callisto and Ganymede formed in just such a manner. Callisto has thirteen crater chains (e.g., figures 13.6 and 13.7); ten are on the side facing Jupiter, and the remaining three are close to that side. The catenae range in length from 200 to 650 kilometers and are composed of up to twenty aligned small craters. Ganymede, which has a younger surface than Callisto, has three crater chains. Europa and Io, with progressively younger surfaces, have no catenae.

Jupiter is not the only planet capable of disrupting comets that come too close. The same thing could happen to every planet in the solar system. The Moon has a crater chain called Davy Catena that consists of about twenty aligned small impact craters (figure 13.8). It seems likely that they were formed by an SL9-like comet that came close to Earth (probably not much farther than

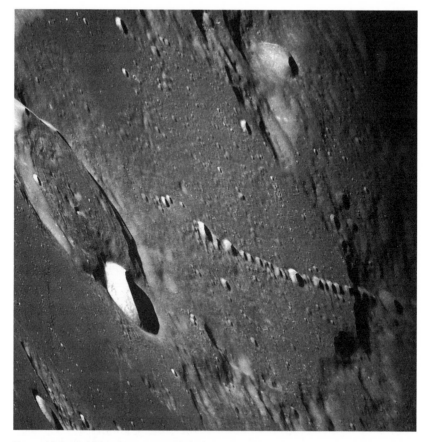

Figure 13.8. A 48-km-long crater chain known as Davy Catena on the Moon located at 11°S, 7°W. The chain occurs within an old, ill-defined, lava-flooded crater known as Davy Y. The large crater at left is Davy Crater (35 km in diameter), named after the English chemist and physicist Sir Humphry Davy (1778–1829). The small crater atop Davy Crater's rim is Davy A (15 km). (Courtesy of NASA)

the 1972 fireball over Montana) and was disrupted by tidal forces. It slammed into the Moon a day or two later. From time to time other SL9-like comets must have struck Earth.

As we have seen from the twentieth-century impacts and near misses, collisions are a fundamental and ongoing phenomenon in the solar system. In order to avoid potential disasters in the near-term, we must first identify the Earth-crossing asteroids and keep an eye out for comets.

Because energetic impacts have occurred on Earth in the distant past, a useful way to study their effects is to examine the crater ejecta. Among the materials thrown out of some craters during major impacts are small glassy objects known as tektites. In some cases tektites and other impact glasses are the only pieces of evidence we possess of major impacts on Earth. The nature of these fascinating rocks and the realization that they were produced by impacts are explored in the next chapter.

XIV

TEKTITES:
A GLASS MENAGERIE

O

It's no go my honey love, it's no go my poppet;
Work your hands from day to day, the winds will blow the profit.
The glass is falling hour by hour, the glass will fall forever,
But if you break the bloody glass you won't hold up the weather.
  —Louis MacNeice, *Bagpipe Music*, last stanza

"PROBABLY the most frustrating stones on Earth."
This was researcher Henry Faul's characterization of the silica-rich, strangely sculptured objects of dark glass known as tektites. The origin of tektites has been the source of consternation and controversy among scientists for two centuries. The word *tektite* itself (from the Greek *tektos*, meaning "melted") was coined in 1900 by Franz E. Suess, who brought these objects to the attention of the scientific world after describing them as glass meteorites. However, Charles Darwin had described some Australian tektites in 1844, calling them a special type of "obsidianite volcanic bomb." As early as 1788 one worker had considered them to be fragments of glassy lava.

The first known written account of tektites appeared around 950 C.E. in a book by Liu Sun, who described them as black stones possessing a "lovely brilliant luster" and making a ringing sound when struck. In prehistoric times these stones were probably used as charms and amulets. They were used as tools by the Australian aborigines and placed in temples by the Indochinese.

Today, tektites are found in four main localities widely separated throughout the world: Australasia, Ivory Coast, the Czech Republic, and North America. Each of these four strewn fields represents the debris from a separate high-energy impact event.

The Australasian tektites have been picked up in sedimentary materials from Australia, Tasmania, Thailand, Cambodia, Laos, China, the Philippines, the Isle of Billiton near Sumatra, and various other sites. (It is customary to derive subgroup names of tektites from the localities in which they are found, hence: australites, billitonites, indochinites, philippinites, etc.) The Australasian strewn field is by far the largest of the four, covering more than 10 percent of Earth's surface. It extends from south of Japan past Madagascar to the eastern coast of Africa, then across the Indian Ocean to a point somewhere between southern Australia and the coast of Wilkes Land, Antarctica.

These tektites show pronounced aerodynamic sculpturing and primary splash forms; teardrops, dumbbells, ellipsoids, spheroids, and lenses are just a few of the numerous shapes assumed by the tektites (figure 14.1). Australite buttons (figure 14.2) have gone through two episodes of liquefaction: The first episode formed a melt from the parental material during the impact event that ejected the melt from the crater; this melt supercooled into a glass spheroid. The second episode occurred during atmospheric reentry when friction due to air resistance slowed the glass spheroids and liquefied the anterior surface. The melt flowed toward the posterior surface and froze into a flange. The undeniable signs of atmospheric penetration in the flanged buttons have given rise to a plethora of theories of extraterrestrial tektite origin. Proposed hypotheses have included glass meteorites, lunar volcanic ejecta, and cometary debris. One worker suggested in 1938 that tektites

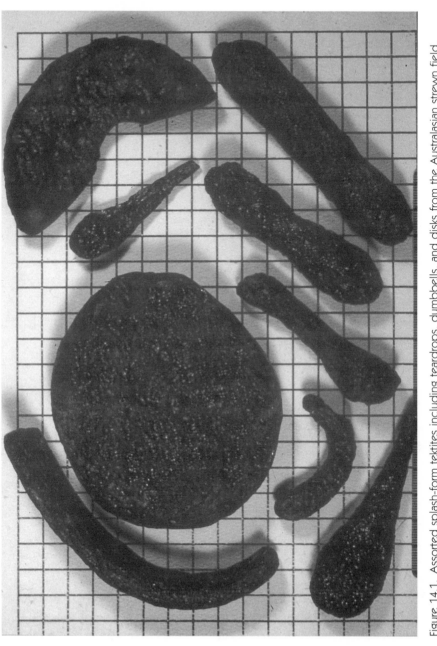

Figure 14.1. Assorted splash-form tektites including teardrops, dumbbells, and disks from the Australasian strewn field. Grid beneath the tektites is divided into square centimeters. (Courtesy of John Wasson, UCLA)

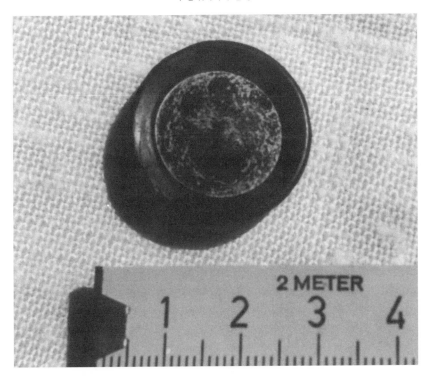

Figure 14.2. Australite button tektite. Scale is in centimeters. (Courtesy of Walter Zeitschel and *Meteorite* magazine)

were derived from nebulous material escaping from large solar prominences.

The Australasian tektites are generally black, although when thin sections are viewed in transmitted light, some are gray, dark brown, or bottle green. Most are a few centimeters in size and are covered with gouges, grooves, and pits formed during terrestrial weathering. They contain no grains that crystallized from the melt (i.e., they are pure glass) and only very rare relict (preexisting) minerals.

In addition to the splash-form tektites, there are large layered tektites, sometimes called Muong-Nong tektites after the type locality in Laos. These objects are blocky chunks ranging up to 32 kilograms in mass. They are distributed across the strewn field

Figure 14.3. Thin slice of a layered tektite from the Australasian strewn field showing folded layers. (Courtesy of John Wasson, UCLA)

between Hainan Island, China, and southern Vietnam. The layered tektites are characterized by dark streaks and zones that have high concentrations of bubbles (figure 14.3). They show prominent layers and appear to have formed by one viscous fluid flowing atop another. Some of the layers are folded. The layered tektites contain inclusions of corundum (Al_2O_3), rutile (TiO_2), quartz (SiO_2), zircon ($ZrSiO_4$), chromite ($FeCr_2O_4$), and cristobalite (SiO_2). In 1993 a 130-gram layered tektite was found in Georgia; its age is approximately the same as that of the other North American tektites.

At the other end of the size spectrum are the microtektites—spheres with diameters less than 1 millimeter. These tiny objects have been recovered from deep-sea sediment cores. Those related to the Australasian tektites have been dredged up from the Indian Ocean, the Philippine Sea, and the western Pacific; those related to the North American tektites have been recovered from the

Gulf of Mexico, Caribbean Sea, and the equatorial Pacific; and those related to the Ivory Coast tektites have turned up in the eastern equatorial Atlantic. Geochemical studies have shown that the microtektites are similar in composition to the normal-size tektites from the adjacent strewn fields. Recently, M. Shyam Prasad and M. Sudhakar reported the discovery of minitektites, 1–3.75 millimeters long, among the Australasian microtektites in the Indian Ocean.

The Australasian tektites contain 62–77 percent silica (SiO_2), 9–18 percent alumina (Al_2O_3), 4–9 percent iron oxide (FeO), 1.3–8 percent magnesium oxide (MgO), 1.3–10 percent calcium oxide (CaO), and 0.5–2.6 percent of the oxides of potassium (K_2O), sodium (Na_2O), and titanium (TiO_2). They are very dry, containing only about 0.01 percent water, far less than in the volcanic glass obsidian (typically about 0.3 percent, but ranging to more than 1 percent). In contrast, the layered tektites are relatively enriched in such volatile elements as chlorine, boron, sulfur, zinc, arsenic, copper, and antimony. Modern argon 40/argon 39 isotopic dating has yielded an age of 770,000 $\pm$ 20,000 years for the Australasian tektites.

The second major group of tektites comes from the Ivory Coast. These dark, olive-green objects were first found near the village of Akakoumoékrou on the banks of the Komoé River. Similar tektites were later discovered in other parts of the country and as microtektites in deep-sea sediment cores taken near the Ivory Coast. Isotopic dating has revealed that these tektites are 1.09 million years old. They contain spherical bubbles 0.1 millimeters in diameter and about 1 percent (by volume) particles of lechatelierite (a silica glass named after the noted French inorganic chemist Henri Louis Le Châtelier). Although the Ivory Coast tektites are quite similar in composition to the Australasian tektites, their tektite-to-tektite compositional variation is far less.

The tektites from the Czech Republic come from Bohemia and Moravia; they are known as moldavites after the Moldau River and represent the only known strewn field in Europe (figure 14.4). Recently, moldavites have also been found in Germany and Aus-

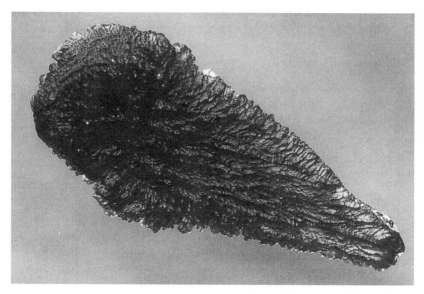

Figure 14.4. A moldavite tektite from 'Chlum nad Malsi in the Czech Republic. (Photograph by Guy Heinen, courtesy of *Meteorite* magazine)

tria. Austrian geochemist Christian Koeberl recently proposed that the name of the strewn field be changed to the "Central European strewn field." Moldavites were the first tektites mentioned in the scientific literature (by Joseph Mayer in 1788) and, like all of the other tektites, were found associated with gravels and other sedimentary deposits. Modern isotopic and fission-track analyses yield an age of about 15 million years for these tektites.

Verification of a North American tektite in 1936 left Antarctica and South America as the only continents on which no tektites have been found. The North American tektites number in the thousands; those found in Texas are sometimes called bediasites (after the town of Bedias, Texas, and the Bedel Indians who formerly inhabited the region around Grimes County, Texas, where they were first found). About thirteen hundred tektites from Georgia (georgiaites) were subsequently recovered as well as a single tektite from the island of Martha's Vineyard off the coast of Massachusetts in 1959 and a single tektite from Cuba around 1970. Most researchers consider the Martha's Vineyard tektite to have been brought there by visitors but regard the Cuban tektite as a genuine in situ find. Both of these specimens are similar in composition to other North American tektites. Deep-sea-deposited microtektites were found in the Caribbean Sea and on Barbados (in an area that was underwater when they were deposited but was later tectonically uplifted).

The North American tektites have been dated at 35.4 million years. The wide distribution of the North American tektites and microtektites suggests that the original amount of material in the North American strewn field may have been appreciably greater than that of the Australasian field. The comparative scarcity of the North American tektites at present is probably a selection effect of their greater age, most of the specimens having long since been swept from the geologic record.

Cosmic-ray exposure ages of meteorites have been used to determine the amount of time meteorites spent in space as discrete 1-meter-size bodies, but tektites lack the telltale isotopes produced in interplanetary space by cosmic rays. The inference is that tektites could not have spent more than a short time in space, quite possibly no time at all. These results restrict the theories of tektite provenance to the Earth-Moon system. If the tektites had been derived from a body in the asteroid belt (as was suggested by one researcher), the tektites' cosmic-ray exposure ages would be similar to those of meteorites, tens of thousands to tens of millions of years.

There are two major groups of theories propounding a lunar origin for the tektites: ejecta from lunar volcanoes and spallation from meteoritic impact on the Moon's surface. The first of these hypotheses is the more easily dismissed. If indeed tektites were derived from lunar volcanoes, there should be numerous specimens of tektites containing angular fragments of rock and glass shards in our collections. None have been found. Tektite expert Virgil E. Barnes pointed out that any lunar volcanoes forceful enough to project tektites all the way to Earth would have spread tektite ash over the Moon's entire surface. Some of this ash should then have been collected by the Apollo astronauts. It was not.

Photographs from Ranger, Orbiter, and Apollo spacecraft have revealed few features on the lunar surface, other than the flood-basalt-covered maria, that can reasonably be interpreted as volcanic. (The Moon's Marius Hills do seem similar to small shield volcanoes on Earth, but these lunar volcanoes are far too small, and have been dormant far too long, to have given birth to the tektites.) Furthermore, hypothetical lunar-derived tektites falling onto a rotating Earth would be spread over a large geographic region. Some material would probably miss Earth on the first pass and go into a solar orbit that caused it to subsequently collide with Earth. The end result would be tektites spanning a far larger geographic area than is observed. In fact, geochemist Harold C. Urey stated in 1962 that there was no probable way that the tektites could have come from the Moon and landed on Earth with the observed distribution of the Australasian tektites.

Finally, moon rocks brought back to Earth by the Luna spacecraft and the Apollo astronauts bear little chemical resemblance to any known tektite. Using data from preliminary analyses, John A. O'Keefe reported finding tektite glass in an *Apollo 12* sample, but this result was refuted by Elbert King and coworkers after more extensive investigations.

Imagining similarities between tektites and moon rocks, Harvey H. Nininger proposed in 1936 that tektites actually represented lunar material blasted off the Moon's surface by meteoritic impacts. Because the Moon lacks an atmosphere, even small meteorites will reach the lunar surface traveling at least several kilometers per second. Escape velocity from the Moon is low, only about 2.4 kilometers per second; thus, some debris reaches lunar escape velocity during energetic impacts. Some of these fragments will land on Earth as lunar meteorites. Lunar meteorites were long anticipated, but the first one was not recognized until 1985. (As of this writing more than a dozen are known.) It was Nininger's contention and that of some later researchers that tektites represented the missing lunar material. In 1971 aerodynamicist Dean Chapman concluded that the Australasian tektite strewn field was probably splashed out of the lunar crater Tycho along a trajectory coincident with one of this crater's most prominent rays.

In 1969 S. Ross Taylor and Maureen Kaye realized that certain terrestrial sandstones (graywackes, subgraywackes, and arkoses) and loess (siliceous windblown silt) bear striking similarities to tektites in both major and trace element abundances and interelement variations. They also concluded that a lunar source for the tektites was not compatible with lunar surface chemistry as revealed by the unmanned Luna and Surveyor spacecraft. Although their paper appeared before the *Apollo 11* astronauts returned the first actual samples from the Moon, subsequent analyses have verified their conclusions.

The tektites' negligible cosmic-ray exposure ages, chemical incompatibility with lunar material, and close compositional similarity with specific wind- and water-deposited sediments all indicate that tektites are definitely terrestrial. However, it was the analysis of inclusions in the tektites that finally discredited theories of extraterrestrial origin.

The silica glass lechatelierite is found in some desert areas after lightning strikes quartz sand. The heat from this event fuses the sand particles, which then rapidly solidify to form a thin tube of silica glass. These tubes, known as fulgurites, can extend several meters in length. The discovery of lechatelierite particles in tektites indicated that the tektites formed at very high temperatures and prompted Virgil Barnes's 1940 suggestion that tektites were nothing more than terrestrial fulgurites. A major flaw with this hypothesis is that no tektites have ever been found in the Great Plains region of the United States, an area with sufficient quartz sand and summer lightning to produce numerous fulgurites. None of the fulgurites from this area bear even a slight resemblance to any known tektites. Also, the mineralogical composition of fulgurite specimens is related to the surrounding terrain. Tektite compositions, however, generally do not reflect the chemical characteristics of their site of recovery. These difficulties caused Barnes to abandon the fulgurite hypothesis.

Lechatelierite found on the rim of Meteor Crater formed during the tremendous heat generated by the impact of a huge meteorite with terrestrial sandstone. Silica glass has also been found in other impact structures. The occurrence of lechatelierite in tektites and in meteorite craters led Leonard J. Spencer to suggest in 1933 that tektites were actually terrestrial impactites, material formed on Earth after the impact of a cosmic body. Tiny spherules of nickel-iron are often found in silica glass surrounding these craters, and the discovery of these metallic particles in tektites helped strengthen the impact hypothesis.

The discovery of coesite (a high-pressure form of silica) in indochinites by Louis S. Walter in 1965 served as an additional link between tektites and terrestrial impact craters. A high-temperature phase of zirconium dioxide (ZrO_2), known as baddeleyite, had previously been shown to be the product of impact metamorphism of the mineral zircon ($ZrSiO_4$), when, in 1963, it was discovered as an inclusion in the Martha's Vineyard tektite. Three years later another baddeleyite inclusion was found, this time in a tektite from Georgia. Numerous grains of quartz were discovered in a few indochinites by Virgil Barnes in 1963; these grains represent incompletely melted terrestrial sandy material incorporated in the still-molten tektites shortly after impact. The discovery in tektites of these inclusions was compelling evidence that tektites were a type of terrestrial impact glass.

The very low water content, the presence of highly evacuated bubbles, and the high ferrous (Fe^{2+}) to ferric (Fe^{3+}) iron ratio in the tektites could also be explained by impacts. The extreme temperatures associated with impact would drive off most of the water, leaving the tektites virtually anhydrous. Virgil Barnes and Richard Russell demonstrated that many tektite bubbles originally contained water vapor that later was absorbed by the surrounding tektite glass, leaving a near perfect vacuum. They showed that the traces of water remaining in the tektites were more than sufficient to account for absorption from all tektite bubbles. Data from the steel industry showed that the ferrous to ferric iron ratio in silicate melts exposed to air increases directly with temperature. The ratios observed in tektites are consistent with impact temperatures.

A major question remained: if tektites were formed by the impact of giant meteorites with Earth's surface, where are the associated craters?

The Nordlinger-Ries Basin in southern Germany was proved to be an impact structure in 1961 after Edward C. T. Chao and Eugene M. Shoemaker discovered coesite and lechatelierite in breccia fragments at the site. They were inspired in their search by the discovery a year earlier of coesite at Meteor Crater. Although an impact origin had been suggested for the Ries Basin as early as 1904, confirmation had to await more detailed petrographic and field data. Potassium-argon dating revealed that this 24-kilometer-wide crater is approximately fifteen million years old, identical in age to the moldavites. This prompted the suggestion that the Ries crater was the source of the moldavites. The distance between the center of this crater and the location of the closest known occurrence of a moldavite is about 250 kilometers, implying that the tektites must have been jetted a considerable distance downrange.

The Bosumtwi Crater in Ghana, a lake-filled depression 10.5 kilometers in diameter, has been shown to be the source for the Ivory Coast tektites. Discovery of metallic spherules and coesite associated with this crater demonstrated its impact nature. Subsequent potassium-argon and fission-track analyses indicated that, within the range of experimental uncertainty, the crater's age was approximately one million years (consistent with that of the Ivory Coast tektites). Strontium/rubidium isotopic data clearly illustrated that the parental material of the tektites is the same age as the rocks around the Bosumtwi Crater. The distance between this crater and its tektite strewn field is similar to the distance between Ries Crater and the moldavites, about 250 kilometers.

In the mid-1990s, a 90-kilometer-diameter impact crater was identified beneath the lower Chesapeake Bay, encompassing part of the Virginia coast. The center of the crater is near the town of

Cape Charles on the Delmarva Peninsula. The age of the crater was dated at about 35.5 million years, approximately the same age as the North American tektites. University of Vienna geochemist Christian Koeberl and colleagues suggested in 1996 that the Chesapeake Bay crater was the source of the North American tektites.

The associations between tektite strewn fields and impact craters convinced nearly all tektite researchers that tektites were indeed terrestrial. However, no known crater is associated with the Australasian tektites, the youngest of the strewn fields. Over the years, various craters have been offered as candidates for the Australasian tektites including a hypothetical crater buried under 2.5 kilometers of ice in Wilkes Land, Antarctica, an 18-kilometer-diameter crater named El'gygytgyn in northeastern Siberia, a hypothetical crater buried beneath thick sediment in the Mekong Delta, and another hypothetical 50- to 100-kilometer crater beneath Lake Tonle Sap in Cambodia. However, no Antarctic crater has been found, the El'gygytgyn Crater is the wrong age, and neither craters nor shock-metamorphosed rocks have yet been found near the Mekong Delta or at Tonle Sap.

A possible reason that a large crater has not been found to be associated with the Australasian tektites despite their young age is that there isn't one. Geochemist John Wasson pointed out that the large layered tektites extend along a region at least 1,140 kilometers in extent. This would require a crater situated near the center of this field to have been so large that the impact could have produced sufficient energy to throw 10-kilogram chunks of ejecta (the size of many layered tektites) distances up to 570 kilometers. It seems unlikely that such a large, young crater would remain hidden.

Virgil Barnes and Kaset Pitakpaivan showed in 1962 that the layered tektites had formed by the flowing of viscous melts. Wasson suggested in 1991 that the melt that formed the layered tektites rained out of the sky as fused dust grains directly below the fireball produced during an impact event. Because the radii of fireballs are equivalent to only a few crater radii and the layered tektites are scattered over a wide region, Wasson proposed that a large number of kilometer-size craters formed in the region. Small craters such as those hypothesized by Wasson would be difficult to locate in highly vegetated areas. Although the failure to locate them thus far may not be too surprising, their discovery would support Wasson's hypothesis.

The projectiles that formed these craters would have been traveling together as a swarm of objects. Wasson postulated that the hypothetical swarm consisted of gravitationally bound remnants of a comet that broke apart in deep space prior to collision with Earth. The analogy was with Comet Shoemaker-Levy 9, which plunged into Jupiter as a series of twenty-odd projectiles in July 1994.

A type of material clearly related to tektites is Libyan Desert Glass (figure 14.5). This material occurs as chunks of glass ranging from millimeter-size particles to 26-kilogram blocks. These chunks are dispersed over a wide area (about 6,500 square kilometers) in the Great Sand Sea in western Egypt, near the border with Libya. They have layers of bubbles, submillimeter-size inclusions of cristobalite (a high-temperature silica mineral) containing small grains of iron-aluminum oxides, rare inclusions of lechatelierite and baddeleyite, and rare needle-shaped clusters of an aluminosilicate mineral (possibly mullite). Also present are dark streaks of iron-rich glass. Libyan Desert Glass closely resembles layered tektites in structure but is much lighter in color and far richer in silica (about 98 percent by weight vs. about 65–75 percent). It formed about 29 million years ago, most likely during an impact event that fused vast amounts of quartz sand or sandstone, generating a highly viscous liquid at the surface that slowly flowed over the landscape beneath the fireball. No known impact crater is associated with Libyan Desert Glass. It could be buried beneath the sands of the Sahara; or there may never have been a crater if the projectile exploded in the air.

After a consensus had developed that tektites were indeed terrestrial materials, planetary geologist Elbert King lamented this fact by saying that tektites had probably received much more at-

Figure 14.5. A smoothed, rounded 80-gram piece of Libyan Desert Glass. (Photograph by G. Müehle, courtesy of *Meteorite* magazine)

tention than they deserved. But King was wrong. Tektites are the products of impacts at Earth's surface or airbursts just above it. Many aspects of tektite origin are still unresolved, and researchers are intent on solving the puzzle. But tektites reveal their secrets only grudgingly because, to paraphrase Paul in 1 Corinthians 13:12, " we can [only] see through a glass, darkly."

XV

RINGS AND SHEPHERDS

O

'Twas noontide of summer,
And midtime of night,
And stars, in their orbits,
Shone pale, through the light
Of the brighter, cold moon,
'Mid planets her slaves,
Herself in the Heavens,
Her beam on the waves.
—Edgar Allan Poe, *Evening Star*

IMPACTS can do more than produce craters, brecciated rocks, and tektites. They can do more than create a Moon or cause mass extinctions on Earth. Impacts into small moons, orbiting the giant planets can eject debris that collects into rings. Complex gravitational interactions among impact debris, small moons and giant planets stabilize the rings. These interactions can result in the most remarkable planetary features in the solar system. All the giant planets have rings. Here is their story.

Saturn

Three hundred eighteen years before Walt Disney produced *Steamboat Willie*, Galileo peered into the heavens and saw Mickey Mouse. What startled Galileo that summer night in 1610 when he trained his crude spyglass on Saturn was a glowing planetary globe with "ears." He wrote that "Saturn is not a single star, but three together, which touch each other. With a telescope which magnifies 1,000 times, the three globes can be seen very distinctly, almost touching, with only a small dark space between them." Galileo observed that Saturn's appearance did not change over the course of more than a year. If the "ears" or "handles" had been moons revolving around Saturn as Jupiter's moons revolved around Jupiter, changes in their positions should have been noticeable every few days. Perplexed at Saturn's peculiar appearance and anxious to work on other projects, Galileo stopped observing Saturn for over a year. When he returned to it in late November 1612, he was astonished to find that the triple star had vanished; only a solitary globe remained. "Now what can be said of this strange metamorphosis?" he wrote. "That the two lesser stars have been consumed, in the manner of the sunspots? Has Saturn devoured his children?" Betraying a rare loss of confidence, Galileo wondered whether the ears had been "an illusion and a fraud with which the lenses of my telescope deceived me for so long—and not only me, but many others who have observed it with me?" Nevertheless, Galileo predicted that Saturn's companion stars would return in due course. They did.

Galileo never solved the mystery of Saturn's disappearing-reappearing ears, but the first step toward a solution was made in 1655 when Dutch astronomer Christiaan Huygens discovered Saturn's sole large moon, Titan. Huygens saw that Titan revolved around Saturn every 16 days. From Titan's orbital movements and Saturn's motion against the fixed stars, Huygens concluded that Saturn was tilted more than 20° relative to the ecliptic (the plane of Earth's orbit around the Sun and nearly the same plane as Saturn's orbit). The modern value of Saturn's tilt is 26.73°, not too different from Earth's tilt of 23.45°. Huygens assumed that Saturn's handles were probably in the same plane as Titan's orbit. Because the handles were closer to Saturn than Titan was, they should revolve around the planet in less than 16 days. Huygens reasoned that a broad, flat, homogeneous ring would not change its appearance as it revolved around Saturn. In addition, if the ring were very thin, it would seem to disappear when viewed edge-on from Earth.

Huygens published a pamphlet in 1656 disclosing his discovery of Titan and including an anagram of his interpretation of Saturn's ring. Three years later he published *Systema Saturnium* and included the solution of the anagram: Saturn had a "thin flat ring, nowhere touching [the planet] and inclined to the ecliptic."

In later years, Huygens used the ring hypothesis to predict Saturn's telescopic appearance. As his predictions were verified, astronomers came to accept the idea of Saturn as a ringed planet. Their drawings showed a broad flat ring, just as Huygens had observed. This consensus lasted until 1675, when astronomer Jean Dominique Cassini reported a dark line (now known as Cassini's Division) within the rings, apparently dividing them into two segments — an inner ring and an outer ring.

Most observers agreed that the double rings were probably very thin; otherwise, they would still be visible when viewed edge-on. But the question soon arose of how thin, solid rings could withstand the strain of being so near to Saturn. The dilemma was addressed by Cassini and others, who suggested that the rings were not solid disks but rather consisted of a swarm of small moons or particles, each orbiting independently. Contrary to Cassini's interpretation, the eminent German-English astronomer William Herschel declared the rings to be solid. Pierre-Simon de Laplace modified this suggestion and modeled the rings as a multitude of solid narrow ringlets, one inside the other, each orbiting Saturn at its own rate.

Over the following decades, observers reported a myriad of contradictory observations. Some saw only the Cassini Division;

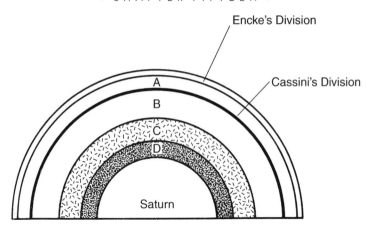

Figure 15.1. Cross section of Saturn and its main rings showing Encke's Division in ring A and Cassini's Division between rings A and B.

some saw two dark lines separating the rings, some three dark lines, others six. One team reported a "great number of narrow rings . . . not unlike a series of waves." The only new feature that most observers recognized was a gap near the center of the outer ring discovered by Johann Encke (and now known as the Encke Division).

In 1850 astronomer William Dawes and the father-and-son team of William and George Bond discovered a faint third ring inside the other two. The new ring was called ring C, the middle ring (the one planetward of the Cassini Division) was called ring B, and the outer ring was called ring A (figure 15.1). In 1852 a few astronomers reported that they could see Saturn shining through the C ring, thereby providing strong evidence against Herschel's idea of solid rings. Scottish physicist James Clerk Maxwell laid the idea of solid rings to rest in 1857 by demonstrating that any slight perturbation (for example, by the gravitational pull of a small moon) in the orbit of a solid ring would cause one side of it to drift closer to the planet. This side would then be tugged on by Saturn to a greater extent than the opposite side of the ring and would drift even closer. Saturn's gravity would accelerate the drift, and the closer part of the ring would crash into the planet.

Maxwell concluded that "the only system of rings which can exist is one composed of an indefinite number of unconnected particles revolving around the planet with different velocities according to their respective distances."

But what causes the gaps in the rings? In 1866 American astronomer Daniel Kirkwood had proposed that gaps in the orbital periods of asteroids were due to resonances with Jupiter. For example, asteroids that orbited the Sun twice for every single time that Jupiter orbited the Sun were in Jupiter's proximity every second orbit. Gravitational tugs by Jupiter modified the orbits of these asteroids into ones of higher eccentricity. This resulted in a paucity of asteroids with these orbital periods. In 1867 Kirkwood proposed that the gaps in Saturn's rings were caused by the same basic mechanism, this time by resonances with Saturn's moons (figure 15.2). For example, any particles that drifted into the Cassini Division would orbit the planet with a period one-half that of Saturn's moon Mimas. The repeated gravitational tugs by Mimas would nudge the particles out of the division.

During the following decades, observers made numerous contradictory reports of gaps in Saturn's rings, many corresponding to resonances with Mimas. Kirkwood had predicted where the gaps should be, and observers, pushing the resolution limits of their telescopes, provided them in great numbers.

Theoreticians turned their attention to why Saturn's rings were so thin. In 1920 British astronomer Sir Harold Jeffreys reasoned that the rings must be vanishingly thin. If Saturn had started out with a spherical swarm of particles orbiting the planet in every direction, mutual collisions would have caused the particles to lose energy. Each particle would tend to reach the average velocity of all the particles. Furthermore, a rapidly rotating gas giant like Saturn has a substantial equatorial bulge that distorts the gravity field and causes the plane of each of the particles to wobble. Particles in inclined or eccentric orbits would tend to collide with other particles; the end result would be for the particles to achieve a circular orbit in the planet's equatorial plane. In the absence of other forces, the ring would be one particle thick.

Figure 15.2. Saturn and moons Enceladus, Dione, and Tethys, photographed by the *Voyager 1* spacecraft from a distance of 106,000,000 km. (Courtesy of NASA)

The rings may have formed from particles around the planet that failed to accrete into a single substantial moon, or they may be debris created from a moon that disrupted. Because the ring particles are presently in coplanar circular orbits, when they occasionally collide, they do so at relatively low velocities. At these low velocities, substantial fragmentation is unlikely, and it seems possible that the particles could stick together and accrete into a moon. The reason that they have not is because in orbits close to a planet, the difference in the gravitational pull by the planet on the nearside and the farside of such a moon would be great enough to tear the moon apart. These gravitational tugs are called tidal forces and although they cannot pull apart small tough objects such as rocks and orbiting spacecraft that are held together by chemical bonds, they can disrupt a liquid moon or a zero-strength conglomeration of rocks and dust bound together only by gravity.

The distance at which the tidal forces will disrupt a liquid moon is approximately 2.5 planetary radii from the center of the planet. Most planetary rings are within this tidal stability limit, which is often called the "Roche Limit" after the French astronomer Édouard Roche, who first calculated this effect.

Further advances in our understanding of the nature of Saturn's rings were made in the late twentieth century. In 1970 Earth-based infrared spectroscopy revealed that the rings were made mainly of water ice. In 1973 radar signals reflected off the ring particles showed that, on average, they are a few tens of centimeters in diameter. Close-up views of the rings were made by *Pioneer 11* in 1979 and by *Voyager 1* and *2* in 1980 and 1981. The rings were imaged in stunning detail (figure 15.3), revealing several surprising features including the observation that some ring particles are at least as big as a house.

Each of the rings was found to be composed of numerous ringlets. Five bands of ringlets flanked by dark gaps occur within the Cassini Division. The Encke Division has unusual ringlets that vary in thickness from one place to another. Although most rings are circular, there is a narrow eccentric ring close to Saturn, in-

Figure 15.3. Saturn's rings, photographed by the *Voyager 1* spacecraft from a distance of 740,000 km. (Courtesy of NASA)

Figure 15.4. Saturn's apparently "braided" F ring, photographed by the *Voyager 1* spacecraft from a distance of 760,000 km. (Courtesy of NASA)

side one of the gaps in the C ring. This ring is composed mainly of small particles revolving around Saturn in the same eccentric orbit. Four thousand kilometers beyond the outermost major ring (the A ring) is the F ring (figure 15.4); this unusual ring varies in thickness from 30 to 500 kilometers and appears to be composed of several small intertwined ringlets. The peculiar braided structure of the F ring is caused by the gravitational tugs of two small moons, Pandora (90 kilometers in diameter) and Prometheus (113 kilometers), one on each side of the ring. Because the moons herd the ring particles and control their movements, they have been likened to two sheepdogs herding sheep and are called shepherding satellites.

A ring particle orbiting Saturn just inside the orbit of a shepherding moon goes faster than the moon and can overtake it. As it closes in on the moon, the moon's gravity accelerates the particle forward. As a consequence of this increased velocity and the law of conservation of angular momentum, the particle moves into a larger orbit. After passing the moon and achieving a larger orbit, the particle is closer to the moon than it was before the encounter. The shepherding moon's gravity tugs on the particle, giving it a backward acceleration; the particle loses energy and thereby falls into a smaller orbit. The net result of these gravitational nudges is the apparently braided structure of the F ring.

Two unusual rings occur beyond the F ring. The outermost ring is the 300,000-kilometer-wide E ring, stretching between the orbit of Mimas at 3 Saturn radii from the planet's center to just past 8 radii. The brightness of the E ring peaks sharply at the orbit of Enceladus at 3.95 Saturn radii. Ring particles consist mainly of micrometer-size blue spheroids; these particles are so small that they would be dispersed by solar radiation pressure in a few tens of thousands of years. The rings therefore must have been produced in the relatively recent past. Because the brightness of the E ring peaks at the orbit of Enceladus, it seems likely that the ring particles were derived from this 500-kilometer-diameter moon. The surface of Enceladus is made of water ice; the low bulk density of the moon (1.2 grams per cubic centimeter) indicates that Enceladus contains little besides water and ice. The lack of craters on Enceladus indicates that the surface is no more than a few hundred million years old and is possibly much younger. It seems plausible that liquid water from Enceladus's interior flowed across the surface in the relatively recent past, erasing all preexisting impact features. This water may also be responsible for the E ring particles. Researchers have speculated that an impact into Enceladus within the past fifty thousand years or so blasted water droplets into space. They would have frozen quickly into ice crystals. The drift of these particles through the Saturnian system produced the broad E ring.

Between the F ring and the E ring lies the G ring, a tenuous

1,000-kilometer-wide band composed primarily of millions of 0.5-millimeter-size particles. Because micrometeoroid bombardment and interactions with plasma in Saturn's magnetosphere would erode the particles in a few thousand years, they must have been injected into this region over this timescale. Although no known moon occurs at this distance (2.8 Saturn radii from the planet's center), a hypothetical source for the G-ring particles would be undetected moonlets, possibly house-size icebergs.

Another surprising discovery in the Voyager images was a set of dark streaks roughly perpendicular to Saturn's rings that resemble spokes on a wheel. The spokes are composed mainly of micrometer or submicrometer dust grains. A probable explanation for the spokes involves exposure to sunlight, which causes the particles to acquire an electric charge; electrostatic repulsion forces the charged particles to levitate off the rings. Scattering of light off the dust grains creates the appearance of dark spokes in the rings.

Uranus

The next ring system to be discovered was that of the planet Uranus. Shortly after his discovery of Uranus in 1781, William Herschel suspected that the planet might have rings. However, the observations were difficult, and he eventually concluded that the rings were not real. This conclusion was a valid one because the real Uranian rings are far too dark for Herschel to have observed them. It seems likely that the distorted image of Uranus he perceived was due to his telescope's astigmatic optics.

It was not until 10 March 1977 that Uranus's rings were first detected. MIT astronomer James Elliot and his team were aboard the Kuiper Airborne Observatory on a C-141 aircraft flying over the southern Indian Ocean. Their aim was to observe a rare occultation of a star by Uranus. The manner in which the star flickered out as Uranus passed in front of it would provide information on the temperature of the Uranian atmosphere; the amount of time the star spent behind Uranus would help determine more precisely than before the planet's diameter and oblateness. Five minutes after the team began monitoring their equipment, the star's light unexpectedly dimmed for a few seconds and then returned to full brightness. The star blinked out a total of five times for periods of two to eight seconds before passing behind the planet. In their recorded conversation, Elliot and his team mentioned several possible explanations for the dimming: narrow rings, debris from ancient Uranian spacecraft, orbiting orange juice cans, and a swarm of space-based honeybees. But what they really thought they had detected was a belt of satellites.

Since that night, the Uranian rings have been studied many times by stellar occultations. Prior to the 1986 *Voyager 2* encounter of the Uranian system, a total of nine dark, narrow rings (most about 10 kilometers wide) had been discovered. Two additional narrow rings and one broad ring were detected by *Voyager 2* (figure 15.5). The outermost ring, called the epsilon ring, is the most eccentric in the system — one side lies 700 kilometers closer to the planet than the other. The ring also varies in width, from 22 kilometers on the side closer to Uranus to 93 kilometers on the farther side. The epsilon ring consists of thousands of small ringlets, each containing a different amount of material. The particles are about 20 centimeters in size and are very dark, similar in color to charcoal. Another peculiar ring near the center of the system is the eta ring; it consists of a 55-kilometer-wide, low-density swath with a higher-density fringe at its inner edge.

In addition to the five approximately 500 to 1,500-kilometer-diameter moons known prior to the *Voyager 2* encounter with Uranus, the spacecraft photographed ten new small moons 26 to 154 kilometers in diameter. Cordelia (26 kilometers) and Ophelia (30 kilometers) are the inner and outer shepherds for the epsilon ring. Additional small moons are near resonances with other rings, and one, Belinda (66 kilometers), is resonant with the gap between rings eta and gamma. Gravitational tugs by these moons confine the ring particles and enable Uranus to maintain its narrow rings.

Figure 15.5. Uranus's rings, photographed by the *Voyager 2* spacecraft. Short star trails are visible in this long-exposure image. (Courtesy of NASA)

Jupiter

The first hints of a ring around Jupiter came in 1974 when *Pioneer 10* and *11* passed through the Jupiter system and detected a few meteoroid hits of marginal statistical significance in the plane of Jupiter's equator. *Pioneer 11* also detected a dip in the radiation intensity of Jupiter's magnetosphere at about 1.8 Jupiter radii from the center of the planet. In hindsight, we know that the meteoroid collisions were probably ring particles, and the decrease in radiation was due to the sweeping up of charged particles by the ring. However, in 1979 the presence of a ring around Jupiter was considered a long shot by most planetary scientists.

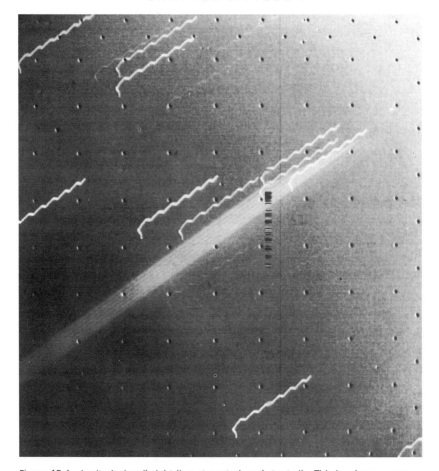

Figure 15.6. Jupiter's ring (bright line at center) and star trails. This is a long-exposure photograph made by the *Voyager 1* spacecraft. Internal spacecraft vibrations caused the wiggles in the star trails. (Courtesy of NASA)

American astronomer Tobias Owen had to convince fellow members of the Voyager imaging team to search for a ring. They agreed to look, and to the surprise of many, they found a ring on 4 March 1979 in the planet's equatorial plane at 1.8 Jupiter radii (figure 15.6).

The Jovian ring system consists of a bright ring less than 30 kilometers thick composed of micrometer-size dust grains, a diffuse 10,000-kilometer-thick halo surrounding the bright ring (consisting of submicrometer grains), and a very faint "gossamer" ring extending out to about 100,000 kilometers beyond the main ring. Because ring particles continuously fall into Jupiter's atmosphere, they must be replenished by some source, most likely a moon.

Voyager 2 images revealed the presence of two small moons: Adrastea (20 kilometers) at 1.80 Jupiter radii, near the outer edge of the ring, and Metis (40 kilometers) at 1.793 radii, within the bright portion of the ring. Analysis of the images of the Jovian ring by the *Galileo* spacecraft in 1999 showed that Adrastea and Metis are the sources for the main ring and dusty halo, and that Thebe (100 kilometers) and Amalthea (195 kilometers) supply particles to the faint outer "gossamer" ring. The collisions of small interplanetary meteoroids on the surface of these bodies produce ejecta that escape the low gravity fields of the moons. The ejected particles drift into stable orbits around Jupiter and form rings.

Neptune

Although English astronomer William Lassell reported a suspected ring around Neptune within two weeks of the planet's discovery in 1846, most observers remained skeptical. No serious attempts to detect a Neptunian ring were made until the 1970s and 1980s, when researchers measured several occultations of stars by Neptune. These studies gave contradictory results: several found no evidence of rings at all, and others showed the star dimming on one side of Neptune but not the other. These results led some planetary scientists to suggest that Neptune was surrounded by discontinuous arcs rather than complete rings. Although many theoreticians doubted that such structures could exist, SUNY researcher Jack J. Lissauer suggested that as-yet-undiscovered shepherding moons could stabilize the arcs. A few workers proposed that the presence of a large satellite in a retrograde orbit relatively close to the planet might preclude the for-

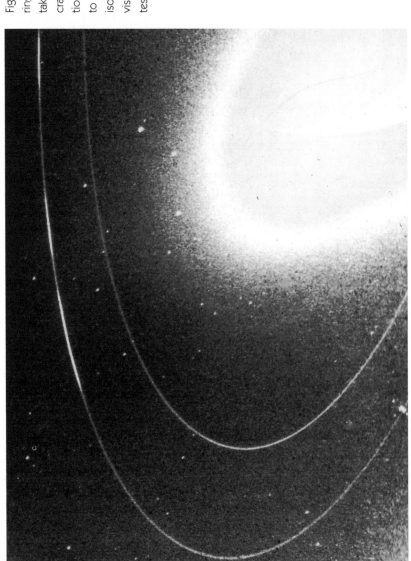

Figure 15.7. Neptune and its rings in a long-exposure image taken by the *Voyager 2* spacecraft. The planet was intentionally overexposed in order to image the dark rings. Three isolated clumps of material are visible in the outer ring. (Courtesy of NASA).

mation of a ring. (Neptune's largest moon, Triton [2705 kilometers in diameter], is in a retrograde orbit with a mean distance of 14.6 Neptunian radii from the planet's center.) With these uncertainties in mind, members of the planetary-science community eagerly awaited the *Voyager 2* encounter with the Neptunian system in August 1989.

Voyager 2 detected five rings around Neptune, including three narrow rings similar to those around Uranus and two dusty rings (figure 15.7). In order of increasing distance from Neptune, the narrow rings were named Galle (after German astronomer Johann Galle, who discovered Neptune telescopically), Leverrier (after French mathematician Urbain Jean Joseph Leverrier, who independently calculated the position of the unknown trans-Uranian planet that was gravitationally perturbing the motion of Uranus), and Adams (after John Couch Adams, the English mathematician who was the first to obtain a solution to Neptune's position). The previously hypothesized arcs turned out to be four separate clumps of material within the thin Adams ring resembling nothing so much as different-size sausages on a string. The arcs extend between 4° and 10° along the ring and range up to 15 kilometers in width. One of the dusty rings lies inside the innermost narrow (Galle) ring and extends inward toward Neptune. The other dusty ring extends outward from the inner rings toward the Adams ring.

Voyager 2 also discovered small moons within the ring system. Among these are Naiad (58 kilometers in diameter), Thalassa (80 kilometers), and Despina (150 kilometers), which occur between the Galle and Leverrier rings, and Galatea (160 kilometers), which occurs just inside the orbit of the Adams ring. Galatea appears to be in resonance with the arcs in the Adams ring, helping to stabilize their high concentrations of material. All of these new moons are relatively dark and may serve as sources of the dark ring particles. Consistent with this hypothesis is the occurrence of a tenuous band of dark material in the same orbit as Galatea; particles in this band may be impact ejecta blasted off the moon but caught in orbit around Neptune.

As we have seen, ring particles are an assortment of impact debris and material freed from tidally disrupted bodies. Gravitational interactions among the particles, shepherding moons, and giant planets form and replenish the rings. Resonances can create gaps in the ring systems as repeated gravitational tugs of planets and moons nudge small ring particles aside. In the words of John Milton, "every shepherd tells his tale." But the complex dynamics of ring particles and the possibility of unknown shepherds around the giant planets make the tale incomplete.

Part 4

LIFE BEYOND EARTH

AS FAR as we know, life in the universe exists only on Earth. Because the observable universe consists of perhaps a hundred billion galaxies, each with billions to trillions of stars, the restriction of living organisms to Earth has seemed inherently implausible to many. Even in past centuries, when the universe was thought to be far smaller, many philosophers believed that they lived in a populated cosmos. At the dawn of the twentieth century, wishful thinking about the existence of extraterrestrial life led many astronomers to perceive evidence of life beyond the resolving power of their analytical instruments. At the beginning of the twenty-first century, some still perceived extraterrestrial life where others did not.

Because Martian meteorites are known to have fallen to Earth, some researchers are exploring the possibility that life may have arisen on Mars and then hitched a ride to Earth within impact ejecta. Alternatively, terrestrial life could have been transferred to Mars by major impact events. This idea of a transfer of life across space from one planet to another is known as panspermia. It is a field of study that ties planetary collisions into the new science of astrobiology. Panspermia also highlights the interconnectedness of the solar system—life that formed on one planet or moon can potentially infect another.

The following chapters on astrobiology provide an excellent example of how science has been practiced in recent times by some of the leading lights in the exploration of the solar system. Many mistakes have been made, and many scientists have succumbed to wishful thinking. But other scientists were always around to challenge provocative claims.

It *is* exciting to speculate on the possibility that there is microscopic life in Martian soil or that organisms are consigned to perpetual darkness in a subglacial ocean on Europa. Excitement can fuel spacecraft missions and command observing time on big telescopes. But we need to remember that we can easily be fooled; positive indications of extraterrestrial life will need to withstand the most careful scrutiny.

In this section we explore the history of the search for life on Mars, analyze the potential for panspermia, reflect on the likely abundance of alien technological civilizations in the Galaxy, and examine our probable responses to the hypothetical detection of a technological extraterrestrial society.

XVI

THE SEARCH FOR
LIFE ON MARS

○

Sorrow is knowledge: they who know the most
Must mourn the deepest o'er the fatal truth,
The Tree of Knowledge is not that of Life.
—Lord Byron, *Manfred*

"ACROSS the gulf of space, minds that are to our minds as ours are to those of the beasts that perish, intellects vast and cool and unsympathetic, regarded this Earth with envious eyes, and slowly and surely drew up their plans against us." When H. G. Wells wrote those words in the serialization of *The War of the Worlds* in 1897 he was capitalizing on a twenty-year history of the life-on-Mars debate. After politics, the weather, and local gossip, no topic so gripped the Western imagination in the late nineteenth and early twentieth centuries as much as whether there were Martians. In Wells's story, when Martian missiles began landing on Earth, they were first thought to be meteorites. And now, more than a century after Wells's invasion, cosmochemists argue about whether or not there are signs of life in meteorites from Mars.

Figure 16.1. The planet Mars, photographed by the Hubble Space Telescope. (Courtesy of the Space Telescope Science Institute)

The history of the search for life on Mars has five main chapters: (1) the hunt for canals, (2) the pursuit of radio signals, (3) the search for vegetation, (4) orbital images of the Martian surface and the Viking Lander experiments, and (5) the study of Martian meteorites. Each episode is filled with colorful characters and heated debate. No other planet has generated as much interest or speculation as Mars (figure 16.1). In the realm of science fiction, men sometimes went to Mars (for example, Edgar Rice Burroughs's series of John Carter novels and Ray Bradbury's *The Martian Chronicles*), and Martians sometimes came to Earth (e.g., Wells's *The War of the Worlds* and Tim Burton's 1996 film *Mars Attacks!*). In the realm of space science, no other planet has been visited by as many spacecraft as Mars, from *Mariner 4* in 1965 to a series of spacecraft in the 1990s (including several failed missions) and *Mars Odyssey* in 2001 at the beginning of the twenty-first century. Although some space scientists are reluctant to admit the real reason for this attention, retired NASA chief Daniel Goldin is not. As he said after the success of the 1997 Mars Pathfinder mission, the "quest for life, this realization that we might not be alone in the universe, is a primary force driving our Mars missions."

The first chapter in the search for life on Mars began in 1877, when Mars and Earth were in opposition (that is, aligned on the same side of the Sun) and especially close. Using a modest 8.5-inch refractor, Giovanni V. Schiaparelli, director of the Brera Observatory in Milan, reported his observations of an extended system of *canali* ("channels") across the surface of Mars. Later that same year, Asaph Hall used the largest refractor in the world, the U.S. Naval Observatory's 26-inch telescope, to discover Phobos and Deimos, the two Martian moons. Although Hall failed to see any canals, Schiaparelli was undeterred. By 1879, at the next Mars opposition, he reported that many of the canals were actually doublets. In 1893 he described Mars as a planet with polar caps of ice and snow and explained that during the northern summer, an ephemeral sea developed around the melting polar cap. Water was transported southward by a network of canals, through which it diffused across the arid planet's surface. Schiaparelli thought it likely that the canals were natural phenomena, but he held open the possibility that they were artificially produced.

Captivated by Schiaparelli's vision, Percival Lowell, an energetic businessman and world traveler, used his considerable wealth to found an observatory in Flagstaff, Arizona. Its purpose was to undertake "an investigation into the condition of life in other worlds, including . . . their habitability by beings like or unlike man." He explained that there was "strong reason to believe that we are on the eve of pretty definite discovery in the matter."

An astronomy enthusiast since he was a child, Lowell would not sit back and let others discover proof of life on Mars. He spent many hours at the telescope himself, contending with Earth's turbulent atmosphere and sketching the martian canals as he perceived them. He publicized his findings in three major books: *Mars* (1895), *Mars and Its Canals* (1906), and *Mars as the Abode of Life* (1908). These books and Lowell's lectures were immensely popular in the United States and Europe. Lowell dedicated his second book to Schiaparelli, heralding him as "the Columbus of a new planetary world." Unlike the cautious Schiaparelli, however, Lowell concluded that Mars was indeed inhabited. His failure to observe clouds in the Martian atmosphere provided him with "a *raison d'être* for the canals. In the absence of spring rains a system of irrigation seems an absolute necessity for Mars if the planet is to support any life upon its great continental areas." Lowell cataloged a total of 183 canals, about four times as many as Schiaparelli had claimed. In his first book, Lowell explained that the canals represented "a network of markings covering the disk precisely counterparting what a system of irrigation would look like." There was also "a set of spots placed where we should expect to find the lands thus artificially fertilized, and behaving as such constructed oases should."

Belief in the existence of Martians was prevalent. In 1900 a wealthy French widow named Clara Goguet Guzman offered a reward of one hundred thousand francs for the first person to communicate successfully with beings of another world "with the exception of Mars." She did not wish to part with her money so easily.

It was only two years after the publication of Lowell's first book that the Martians of H. G. Wells invaded Earth. Each of the Martians was the size of a bear and had "two large dark-coloured eyes" and a mouth "which quivered and panted, and dropped saliva." The monstrous Martians had left their dying planet in a covetous quest for Earth's near-inexhaustible supply of water. It is perhaps no coincidence that Wells was a friend of Lowell's and was persuaded by Lowell's findings.

Although some astronomers also reported observing canals during this period, others did not. The spectrum of opinion was diverse. Great astronomers staked claims on all sides of the issue. E. Walter Maunder of the Royal Greenwich Observatory declared the canals illusory. American astronomer Henry Norris Russell sided with Lowell, and Sir Arthur Eddington, like Schiaparelli, thought the canals to be natural features.

The canal controversy was finally resolved to the satisfaction of many by the dedicated efforts of Eugene M. Antoniadi. Although not well known today, Antoniadi was highly regarded by his contemporaries. As head of the Mars section of the British Astronomical Association, he observed Mars at every opposition with his own 8.5-inch (22 cm) telescope or with the 9.6-inch (24-cm) at the observatory in Juvisy, France. In the fall of 1909 he obtained permission to use the 33-inch (84-cm) refractor at Meudon, France, the largest telescope in Europe. He observed Mars for two months under good seeing conditions and concluded that the canals were actually "complex shadings — the integration of irregular details, too minute to be accessible to our means, as Mr. Maunder argued 15 years ago. . . . The geometrical canal network is an optical illusion; and in its place the great refractor shows myriads of marbled and chequered objective fields, which no artist could ever think of drawing."

Writing in the *Journal of the British Astronomical Association* in the wake of these observations, Maunder declared that people need no longer "occupy their minds with the idea that there were miraculous engineers at work on Mars, and they might sleep quietly in their beds without fear of invasion by the Martians after the fashion that Mr. H. G. Wells had so vividly described."

Ironically, few of the dark markings perceived by Antoniadi correspond to actual features on Mars. Chapter one thus ended the way it began — with difficult observations beyond the resolving power of contemporary telescopes.

Nikola Tesla, a great but enigmatic electrical engineer, physicist, and inventor, opened the second chapter of the search for life on Mars in 1901 when he published an article titled "Talking with the Planets" in *Collier's Weekly*. He had observed peculiar chirping noises two years earlier that had "terrified" him and caused him later to feel that he had been the first person to hear a radio signal from another planet. By 1919 Guglielmo Marconi had speculated on wireless communication with extraterrestrial intelligences and suggested that some "very queer sounds" that he had detected (which resembled scrambled letters in Morse code) might have come from beyond Earth. Articles in the popular press linked the Tesla and Marconi observations to signals from Mars.

Although Marconi professed ignorance in 1920 of the source of the unusual signals, within a year he was convinced that they were messages from Mars. He had received new signals that resembled the letter *v* in Morse code: three dots and a dash. The signals had a wavelength of 150 kilometers, approximately ten times longer than those used in terrestrial transmissions. He had been unable to decipher the meaning of the signals, but their regularity impressed him, and he looked forward to the establishment of two-way interplanetary communication. However, even Tesla was skeptical of these claims; he pointed out that interference among radio waves transmitted by stations on Earth could create very long wavelengths. Modern radio astronomers would categorize the noises that Tesla and Marconi detected as low-frequency electromagnetic waves called whistlers. These waves are caused by terrestrial lightning and move slowly along Earth's magnetic lines of force. Marconi apparently accepted Tesla's explanation for the peculiar signals and maintained radio silence for the rest of his life on the subject of extraterrestrials.

Other plans for detecting radio signals from Mars were announced in 1909 and 1920 by David P. Todd, director of the Observatory of Amherst College. Todd intended to place radio receivers in specially constructed balloons in order to minimize atmospheric interference. Although his balloons never got off the ground, Todd later convinced the U.S. Army and Navy to refrain from transmitting radio signals during the close approach of Mars in August 1924. Armed-service radio operators listened intently for Martian radio signals but heard nothing. Canadian and British radio experts reported some unusual sounds, but few attributed them to intentional broadcasts from Mars. The categorization by the Marconi Company (which by that time had experienced a change in personnel) of all attempts to detect radio signals from Mars as "fantastically absurd" brought chapter two to a close.

The third chapter in the search for life on Mars actually began forty-two years before Percival Lowell was born. In 1813 Honeré Flaugergues, a longtime observer of Mars, found that the Martian polar caps changed shape every spring. He deduced that such seasonal changes probably coincided with the spread of global vegetation. Many such observations of a seasonal "wave of darkening" were made throughout the nineteenth century, and by the 1920s most astronomers accepted these reports as evidence that Mars harbored primitive plant life. No one at that time suggested the currently accepted explanation that the wave of darkening is caused by windblown dust. Earl C. Slipher of Lowell Observatory summed up the contemporary view after the Mars opposition of 1924. "The seasonal date that these dark markings begin to darken . . . the rate and manner of their development, the seasonal date at which they mount to the highest intensity . . . their color and appearance, and in turn the time of their fading out again — all obey the law of change that we should expect of vegetation."

Supporting this notion were the first experimental determinations of Martian surface temperatures by William W. Coblentz and Carl O. Lampland of Lowell Observatory and by Edison Pettit and Seth Nicholson of Mt. Wilson Observatory. They used the newly developed vacuum thermocouple at the focus of their large reflecting telescopes and found average temperatures of about $-10°$ to $-30°C$ for the entire Martian disk. Coblentz and Lampland also analyzed bright and dark regions separately and found temperatures of $-10°$ to $+5°C$ and $+10°$ to $+20°C$, respectively. Coblentz concluded that the high surface temperatures of the dark areas could be explained by the presence of living vegetation.

A contemporary snapshot of the public's acceptance of the possibility of life on Mars can be gleaned from Robert Frost's 1923 poem "A Star in a Stone-Boat" (wherein the "star" is a meteorite and the "stone-boat" is a handcart used for hauling stones). The sixteenth stanza reads:

> Sure that though not a star of death and birth,
> So not to be compared, perhaps, in worth
> To such resorts of life as Mars and Earth, —

Although Frost won a Pulitzer prize for the collection of poems in which this appears, it remained unpopular with critics, some of whom may have harbored suspicions that Frost should have taken the road more traveled.

Spectroscopic observations of the Martian atmosphere conducted in 1925 provided some constraints on the vegetation hypothesis. Walter S. Adams and Charles E. St. John, using the 60-inch (1.5-m) reflector at Mt. Wilson, found concentrations of water vapor and molecular oxygen (O_2) equivalent to 6 percent and 16 percent, respectively, of those over Mt. Wilson. These low values indicated that Mars was a desert, but they did not rule out the existence of drought-resistant vegetation.

Although Adams and Theodore Dunham, using the 100-inch (2.5-m) telescope at Mt. Wilson and improved spectrographic equipment, detected no oxygen or water vapor in the Martian atmosphere in 1933, the vegetation hypothesis was still embraced by a majority of astronomers. As late as 1945 Henry Norris Russell wrote in a popular astronomy book he coauthored that the presence of plants on Mars was more likely than not. He based his conclusion on the belief that Martian life may have originated under clement conditions and then adapted to an increasingly harsh environment by natural evolutionary processes.

In 1947 Gerard P. Kuiper used infrared spectroscopy and the 82-inch (2.1-m) reflector at McDonald Observatory to discover carbon dioxide (CO_2) in the Martian atmosphere. Because carbon dioxide and water are the reactants that plants use in photosynthesis, the detection of CO_2 seemed supportive of Martian vegetation. However, because Vesto M. Slipher, acting director of

Lowell Observatory, had failed to detect chlorophyll in the reflection spectra of the dark regions of Mars back in 1924, Kuiper concluded in 1952 that Mars might be covered with lichens (which lack chlorophyll). He based this conclusion on the colors of the Martian surface, the seasonal changes in color, and an argument by Ernst J. Öpik of the Armagh Observatory that dust storms would have covered the dark regions unless there were renewed growth.

At the 1956 and 1958 Mars oppositions, William M. Sinton of Harvard University used infrared spectroscopy to look for organic molecules. Although the observations pushed the limits of his instruments' capabilities, Sinton was confident of his results. In a 1959 *Science* paper titled "Further Evidence of Vegetation on Mars," he concluded that the observed spectrum fit that of organic molecules, particularly those of plants, very closely.

Although these results were widely publicized, the acclaim was short-lived. By 1965 critics had shown, and Sinton had conceded, that his spectral data had been contaminated by deuterium in Earth's atmosphere. In July of that year, *Mariner 4* flew within 9,780 kilometers of Mars and measured the atmospheric pressure at the surface to be about 5 millibars, two hundred times lower than Earth's. The spacecraft also sent back twenty-two television pictures of a small portion of the Martian surface. The pictures revealed a world with numerous impact craters covering smooth plains. Although there were fewer craters than on the lunar highlands, there were more than on the lunar maria. Four years later, *Mariner 6* and 7 flew by Mars and photographed 20 percent of the planet's surface. More craters were observed, and most planetary scientists judged Mars to be a lifeless, Moonlike world. The vegetation hypothesis had withered.

Nevertheless, the stage had been set for the fourth chapter in the search for life on Mars. Besides impact craters, *Mariner 6* and 7 had photographed a belt of chaotic terrain, apparently formed by the withdrawal of subsurface material, triggering the collapse of overlying rock. The spacecraft had also imaged the featureless terrain of the Hellas basin, and some researchers interpreted this region to have been the site of extensive sedimentation that had obliterated preexisting surface features.

In 1971 *Mariner 9* went into orbit around Mars and mapped virtually the entire planet at a resolution of 1 to 3 kilometers. Ancient cratered plains were seen to be mainly confined to the Southern Hemisphere. Huge volcanoes were observed in several regions, accompanied by volcanic plains covered with lava flows. The Martian crust had been fractured by tectonic stresses; extensive rift zones and channels had been created. But the most exciting results were those demonstrating that liquid water had once flowed across the Martian surface. Many researchers reasoned that where there had once been water, there may have been life.

The indications of immense volumes of Martian water in the past bring up a problem: Where is all of it today? One possibility is that it all escaped. If it did not, there are only two obvious reservoirs — polar ice deposits and subsurface ice (permafrost).

The temperature and pressure of the Martian atmosphere today are too low to allow the presence of liquid water. If you emptied a bucket of water on the surface of Mars, the water would freeze quickly. In order for water to have flowed in the past, the atmospheric pressure must have been much higher. (At low pressures, liquids evaporate quickly.) Many researchers believe that the early higher-pressure Martian atmosphere resembled the present one in being dominated by carbon dioxide. They postulate that most of the atmospheric CO_2 was depleted by the formation of extensive deposits of carbonate rocks. It is less likely, but possible, that much of it was blasted into space by high-energy impact events.

Mariner 9 relayed pictures of wide, meandering channels on Mars that appear to be dried-out river valleys. Although far too small to have been seen by Schiaparelli, these channels (*canali* in Italian) appear to have been the sites of running water. One channel, Nirgal Vallis, has branching tributaries (figure 16.2). Others have teardrop-shaped islands with terracelike flanks that seem to have been scoured by floods (figure 16.3). Another channel, Mangala Vallis, has a drainage pattern similar to that of the Chan-

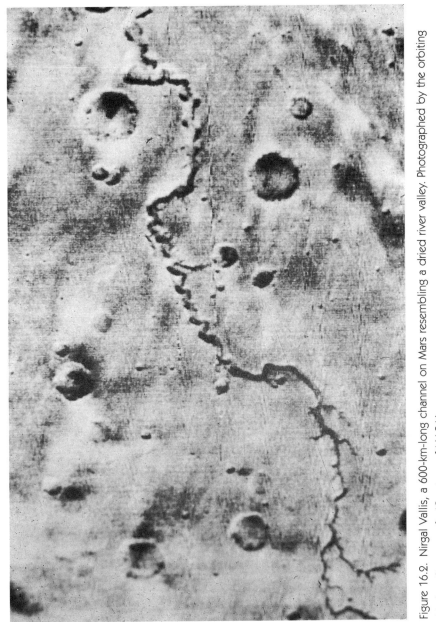

Figure 16.2. Nirgal Vallis, a 600-km-long channel on Mars resembling a dried river valley. Photographed by the orbiting *Mariner 9* spacecraft. (Courtesy of NASA)

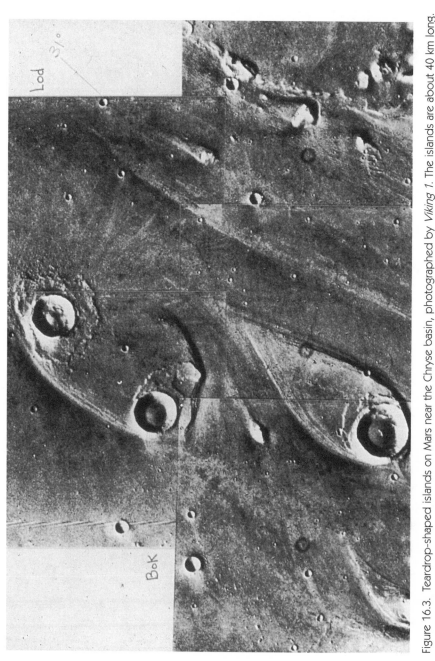

Figure 16.3. Teardrop-shaped islands on Mars near the Chryse basin, photographed by *Viking 1*. The islands are about 40 km long. They formed by erosion of soft rocks or sediments on the lee side of impact craters. The compacted rocks of the impact-crater rims deflected the rushing water. (Courtesy of NASA)

neled Scablands of the northwestern United States. The latter region was the site of a series of immense floods during the Pleistocene when the retaining walls of a glacial lake failed near Missoula, Montana, and huge volumes of water carved out the adjacent landscape.

Martian furrowed terrain consists of irregular sinuous depressions on the flanks of crater rims. These features have been interpreted as landscapes eroded by rainfall. Sinuous channels located downslope of chaotic terrain may have formed by the discharge of water released by melting permafrost. The permafrost may have been melted by volcanic or geothermal activity. Some impact craters have a thick blanket of ejecta that may have contained a substantial amount of water derived from permafrost melted by the impact (figure 16.4).

When did water flow on Mars? The presence of impact craters superposed on some channels indicates that these features are all probably several billion years old.

More than seven thousand pictures were relayed back to Earth before *Mariner 9* ran out of attitude control gas and began to spin slowly out of control. Its transmitter was turned off by radio command in October 1972, and the mission concluded after 698 orbits.

Less than four years later, two Viking spacecraft landed on Mars and ushered in a new era in the search for life. In addition to the television camera, the primary instruments on the Viking landers were three experiments inside the 1-cubic-foot, 15-kilogram biology package. The robotic arm of the lander (figure 16.5) scooped up Martian soil and dropped it into an entry port, where it was distributed among the experiments.

In the labeled release experiment, developed by Gilbert Levin of Biospherics Incorporated, a dilute watery solution of organic compounds labeled with radioactive carbon 14 was added to the soil. Any indigenous Martian microorganisms were expected to behave like their terrestrial counterparts, assimilate the organic compounds, and release waste gases enriched in carbon 14. These would be counted by a detector as disintegrations per minute.

Figure 16.4. The Arandas crater (28 km in diameter) and its thick ejecta blanket. The ejecta may have formed from debris containing substantial amounts of water. The front of the ejecta deposit forms a steep hill rising above the surrounding plain. (Courtesy of NASA)

In the gas-exchange experiment, developed by Vance Oyama of NASA Ames, Martian soil was moistened (in some cases by water containing a rich soup of organic compounds) in the expectation that Martian organisms would increase their metabolic activity and release gas. Gases of different compositions would pass through the absorbing medium above the sample chamber at different rates and be identified by chromatography.

In the pyrolytic release experiment, developed by Norman Horowitz of Caltech, Martian conditions were re-created. Carbon dioxide and carbon monoxide (CO), both containing carbon

Figure 16.5. Robotic arm of the Viking Lander as it extends to scoop up Martian soil. (Courtesy of NASA)

14, were added to the chamber. Assimilation of these gases by indigenous organisms would cause them to release organic matter. Heating of the chamber to 635°C would kill the organisms and release the radioactively tagged organic compounds, which could then be counted.

The fourth biology experiment, although not specifically designated as such, was the television camera. If any reasonably sized plants or animals were in the vicinity of the landers, they would be seen. Carl Sagan said that he once had a nightmare about the Viking spacecraft: every night, little Martian animals would visit the Lander, but all that could be seen by day were their footprints.

In addition to the biology team, there was a molecular analysis team, headed by Klaus Biemann of MIT. His results had grave implications for any Martian biota — no organic molecules at all were present in the soil down to levels of a few parts per billion. Although recent data suggest that the experiment may have been unable to detect heavy indigenous organic molecules, at that time Biemann's results tempered the enthusiasm of the biologists when their experiments seemed to yield positive results. Gases were evolved in the labeled release experiment, but their abundance gradually declined. Gases were also evolved in the pyrolytic release experiment. Carbon dioxide and molecular oxygen were released in the gas exchange experiment, but because oxygen had never been released in similar tests on terrestrial and lunar soils, the results pointed to a chemical reaction, not a biological one. Oyama later concluded that the reaction was caused by highly oxidized components (i.e., peroxides) in the Martian soil.

Horowitz summarized the results in a *Scientific American* article in 1977. "It is impossible to prove that any of the reactions detected by the Viking instruments were not biological in origin. It is equally impossible to prove from any result of the Viking experiments that the rocks seen at the landing sites are not living organisms that happen to look like rocks. Once one abandons Occam's razor the field is open to every fantasy."

Horowitz's statement brought the fourth chapter of the search for life on Mars to a close. Things would have remained there, barring surprise discoveries from future spacecraft, except for startling news published in the journal *Science* in August 1996. David S. McKay of NASA's Johnson Space Center and his coauthors reported evidence for microscopic fossils of bacteria-like organisms in a Martian meteorite called Allan Hills 84001. Their report opened chapter five.

Before discussing McKay's evidence for Martian life, it is important to ask how we know that the meteorite in question (or the seventeen other meteorites currently designated as Martian) really come from Mars. The first clue is the relatively young crystallization ages of most of these rocks (180 million years to 1.3 billion years before present). This eliminates the asteroids, because asteroids are small bodies that cooled completely within 100 million years or so after they formed; volcanic activity on achondritic asteroids had effectively ceased over four billion years ago. Direct evidence for Mars comes from gases trapped within shock-melted glass in one of these rocks (EETA79001) (figure 16.6);

Figure 16.6. Martian meteorite EETA79001. This meteorite has shock-melted inclusions containing gas bubbles of the Martian atmosphere incorporated during its launch off Mars. (Courtesy of NASA)

their abundances and isotopic compositions match those of the Martian atmosphere at ground level measured by the Viking spacecraft in 1976.

Another strong piece of evidence is the high deuterium/hydrogen ratio (D/H) in the samples. Because the surface gravity of Mars is only 38 percent as strong as that of Earth, hydrogen (the lighter of the two gases) escapes from the top of the Martian atmosphere more readily than deuterium. Water in the Martian atmosphere thus has a D/H ratio about five times that of terrestrial water. Phosphate minerals that form from water-bearing fluids on Mars consequently have high D/H ratios compared to those of terrestrial phosphates. The meteorites inferred to be Martian have comparably high D/H ratios in their phosphates.

The Martian meteorites also have distinct oxygen-isotopic compositions, setting them apart from rocks from the Earth, Moon, and known asteroids.

Figure 16.7. Ancient Martian meteorite ALH84001, which some researchers believe contains evidence of fossilized Martian bacteria. (Courtesy of NASA)

The Allan Hills 84001 meteorite was found in Antarctica on the Allan Hills ice field in 1984, but it was not recognized as Martian until NASA scientist David W. Mittlefehldt published a detailed petrologic study of the rock ten years later. Oxygen-isotopic analysis by Robert N. Clayton of the University of Chicago confirmed its Martian pedigree. Unlike the other Martian meteorites, however, ALH84001 (as it is abbreviated) is an ancient rock, four and a half billion years old (figure 16.7). It is the only sample in our possession that can provide key information about the early history of Mars.

In their report on ALH84001, McKay and his team cited four lines of evidence that they believed pointed to probable biological activity: (1) They detected minute amounts (on the order of a part per million) of fused hydrocarbon rings known as polycyclic aromatic hydrocarbons (PAHs), which on Earth can form by geochemical alteration of some hydrocarbons in decaying organisms. (2) Using electron microscopy they imaged tiny grains of magnetite (Fe_3O_4, a magnetic iron oxide) that they found similar in shape to those produced by certain kinds of terrestrial bacteria.

(3) They imaged tiny ovoid and elongated forms (some as small as 0.00002 millimeter in length) within carbonate globules that seemed similar in size and shape to reported occurrences of terrestrial nanobacteria. This length is equivalent to 0.02 micrometers or 20 nanometers (figure 16.8). (4) They interpreted the chemical zoning in the carbonate globules as indicating that the carbonate formed at temperatures low enough to allow life to exist. They commented that "although there are alternative explanations for each of these phenomena taken individually, when they are considered collectively in view of their spatial association, we conclude that they are evidence for primitive life on early Mars."

Given the photogeologic evidence from the *Mariner 9* and Viking Orbiters that water once flowed across the surface of early Mars, it seemed eminently plausible that life may once have been present. Because ALH84001 is the only ancient Martian rock we have, it seemed the perfect place to look. After the McKay team's paper appeared, a number of planetary scientists were comfortable with the notion that we had, at last, found life on Mars.

However, as Carl Sagan occasionally remarked, "extraordinary claims require extraordinary evidence." Harsh criticism of the McKay-group study began almost immediately. Each of their lines of evidence was challenged and, in the opinion of many, refuted.

Several researchers pointed out that PAHs also form by nonbiological processes, and so a biological origin is not demanded. One team measured the PAHs in another Martian meteorite recovered from Antarctica and found that the variety of PAHs in both rocks matched that in Antarctic ice. They suggested that terrestrial contamination was the source of the PAHs. Another research team found that the carbon isotopic compositions of both Antarctic Martian meteorites were within the terrestrial range, consistent with contamination.

Using electron microscopy, John P. Bradley of MVA Incorporated and his colleagues imaged a wide assortment of magnetite morphologies in ALH84001. Crystallographic dislocations in

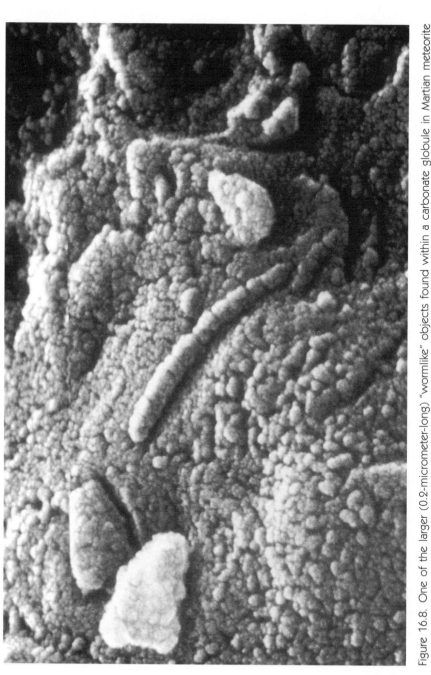

Figure 16.8. One of the larger (0.2-micrometer-long) "wormlike" objects found within a carbonate globule in Martian meteorite ALH84001, as imaged with a scanning electron microscope. This segmented object was interpreted by some researchers as a fossilized Martian bacterium. (Courtesy of NASA)

some of the magnetite grains are unlike those produced biologically and are consistent with growth of the magnetite from a vapor at high temperatures. They also found that many of the magnetite grains nucleated on the carbonate and grew in crystallographic alignment with the carbonate. This rules out biological deposition of the magnetite and is again consistent with growth from a vapor.

Bradley's argument was countered by the McKay team. They studied several hundred magnetite grains in ALH84001 and pointed out that 27 percent of the grains were shaped like elongated prisms and were indistinguishable in five of six characteristics from magnetite grains produced by a terrestrial magnetotactic bacterial strain called MV-1. They also reported that some magnetite grains in ALH84001 carbonate globules occurred in chains surrounded by a membrane. Some species of terrestrial bacteria make similar structures.

The apparent microfossils in ALH84001 are far smaller than terrestrial microorganisms. In fact, the smallest Martian "microbes" are more than two thousand times smaller than bacteria of the genus *Mycoplasma*, the smallest known living terrestrial organism. This organism is itself a parasite living in the cells of larger organisms and is descended from a larger, nonparasitic bacterium. Although McKay and his team claimed that the Martian microfossils are similar in size to terrestrial nanobacteria, the existence of such tiny terrestrial organisms is in dispute. The smallest specimens of the purported Martian microfossils are barely wide enough to hold a few strands of DNA. Some researchers have questioned whether an organism that small could accomplish all the functions required of a living cell. Bradley and coworkers carefully examined the carbonate in ALH84001 and could find no tiny particles other than magnetite that are the same size and shape as the purported microfossils. They concluded that the microfossils are, in fact, magnetite grains. Nevertheless, French researchers led by P. Gillet reported 100-nanometer-size (0.1-micrometer-size) terrestrial bacteria that had invaded fractures in the Tatahouine diogenite meteorite since its fall in 1931. This reopened the question of how small viable organisms could be.

Many groups have continued to perform petrologic, chemical, and isotopic analyses of the carbonates in ALH84001 and arrived at disparate conclusions. Some teams argue that the carbonate precipitated from water- and CO_2-rich fluids at low temperatures; others argue that it formed at high temperatures by volcanic activity or impact-melting. The formation temperature of the carbonate is presently unresolved, although most researchers presently favor formation temperatures less than 400°C.

The McKay team answered many of the criticisms. They proposed that some of the elongated nanofossils are small because they may be dwarf bacteria or detached bacterial appendages. McKay and his colleagues also described delicate lacy networks attached to pyroxene and carbonate grains and suggested that these structures might be analogous to organic secretions of terrestrial microorganisms (biofilms). Skeptics pointed out that the lacy structures have not yet been shown to be organic. The dialogue and the research continue.

In 1999 McKay and coworkers reported the presence of bacteria-shaped objects in another Martian meteorite, Nakhla. Nakhla is much younger than ALH84001. It crystallized 1.3 billion years ago, almost certainly after running water had disappeared from the Martian surface. Putative Martian life could not have invaded this rock until after it had crystallized and cooled. If there was Martian life 1.3 billion years ago, it might still be around today. Planetary conditions have not changed that much since that time. However, most investigators believe that the available evidence does not support the case for Martian life. Some have likened the adherents of Martian life to desperately thirsty desert nomads chasing a mirage.

But it is far too soon to give up the search. Within the last few years, many biologists have concluded that the most primitive *terrestrial* organisms are microbes that live within sedimentary rocks at depths of several kilometers or near volcanic vents on the ocean floor. These organisms thrive at high temperatures (70–

121°C or 158–250°F) and relatively high pressures. They have been dubbed "thermophiles" (i.e., heat lovers) and "hyperthermophiles." They are not dependent on sunlight or oxygen and instead obtain energy from readily available inorganic chemicals such as hydrogen, sulfur, hydrogen sulfide, and iron sulfide. Ancestors of these microorganisms would have been shielded from many of the effects of asteroid collisions, climate change, and intense ultraviolet light that would have plagued surface dwellers. Because such organisms evolved on Earth, it is possible that they evolved on Mars as well, perhaps near hot springs associated with Martian volcanoes. As volcanic activity on Mars tapered off, the number of hot springs was reduced. Fossils of Martian microorganisms might be found in dried-up hot springs near extinct volcanoes. Living Martian microorganisms could still be thriving in isolated locations near still-extant hot springs. Some fossils or dormant microbes could have been excavated by asteroidal impacts and deposited on the surface.

Similarly, because metabolically active bacteria have been found at the South Pole on Earth, thriving in subfreezing conditions, it does not seem out of the question that microbes might exist at the surface of Mars. Such hypothetical microorganisms could conceivably have evolved from thermophilic ancestors.

In June 2000 NASA scientists announced that the *Mars Global Surveyor* had photographed a few hundred gullies on the sloping walls of craters and valleys, mainly restricted to Mars's Southern Hemisphere. The uppermost portions of the gullies appear to have collapsed; the lowermost portions contain debris that was deposited downslope. The gullies appear to be geologically young, at most a few million years old, and possibly far younger. The features were modeled as having been formed by flowing water that broke through small ice dams and sent floods down the gullies. The source of the water was assumed to be a few hundred meters below the surface. If liquid water still exists in favorable locations on Mars, the possibility of present-day Martian microorganisms cannot be discounted.

If life is ever found on Mars, we will need to examine the question of whether terrestrial meteorites brought life there or whether Martian meteorites ever brought life here. Possible transfers of microorganisms constitute another potential connection among solar-system bodies. The diversity of ideas on such transfers is the topic of the next chapter.

XVII

PANSPERMIA

O

A fire-mist and a planet,
  A crystal and a cell,
A jelly-fish and a saurian,
  And caves where the cave-men dwell;
Then a sense of law and beauty
  And a face turned from the clod—
Some call it Evolution,
  And others call it God.
  —William Carruth, *Each in His Own Tongue*

WE WALK within a sea of life. Bacteria, tiny seeds, and fungal spores are borne by the wind and dispersed around the globe. They are joined by dust-size eggs from brine shrimp, infant spiders clinging to silk threads, and hordes of tiny insects. Although few places on Earth are as sterile as a new volcanic island rising above the water, the volcano is erupting within this sea of life. The first spores and seeds burn on contact with the lava, but as the lava cools into rock, more seeds and spores are deposited. Waves carry larger seeds to the shore. Insects, spiders, and lizards raft in on driftwood, tangled roots, and mats of vegetation. Migrating birds stop over to rest and eat; some eventually take up residence. The once-sterile island becomes infected by life. This colonization of islands by organisms drifting in from afar documents the hypothesis of "panspermia," the notion that seeds of life are everywhere. It demonstrates the interconnectedness of the planet.

Since the term was coined by the Greek philosopher Anaxagoras in the sixth century B.C.E., panspermia has been invoked numerous times to explain the origin of life on Earth. Philosophers and scientists drawn to this idea have connected Earth to the cosmos. The history of ideas encompassed by panspermia is a fascinating one. It includes spores drifting in from interstellar space, infected comets impacting Earth, putrescent trash left by littering space-aliens, and bacteria intentionally introduced from incubators on robotic alien spacecraft. Panspermia also serves as the theme of many great and not so great science fiction stories. It is an arena where science, science fiction, philosophy, informed speculation, and theology meet.

In the eighth or ninth century B.C.E., the writer of Genesis 2:7 declared that "the Lord God formed man of the dust of the ground, and breathed into his nostrils the breath of life; and man became a living soul." In Hebrew the word *Adam* is derived from the word "man," but it is also related to the word *adamah*, meaning "ground" or "earth." This passage reflects the ancient belief that life arose from dust after an encounter with an invigorating spiritual force.

A few hundred years later, Anaxagoras expressed a similar thought—that organisms were formed from slimy earth after it was fertilized by invisible, spiritual germs of life. These ideas persisted for millennia. Eight hundred years after Anaxagoras, St. Augustine proposed that, under favorable conditions, widespread invisible, spiritual seeds generated birds, insects, plants and frogs from earth, air, and water.

Not all philosophers proposed spiritual seeds for vitalization. The sixth-century B.C.E. philosopher Anaximander speculated that the first animals developed from sea slime. Two centuries later, Aristotle extended this doctrine of spontaneous generation and described how fireflies arose from morning dew. Later workers claimed that worms formed from mud, maggots from decaying meat, and mice from dirty underwear and wheat. In Shakespeare's 1606 play *Antony and Cleopatra*, Lepidus tells Mark Antony, "[y]our serpent of Egypt is bred now of your mud by the operation of your Sun: so is your crocodile." By the time Antonie van Leeuwenhoek invented the microscope in 1683 and discovered microorganisms in stagnant water, it seemed plausible to many that the peculiar "animalcules" arose spontaneously from nonliving matter.

Yet the theory of spontaneous generation had its critics. A Tuscan physician, Francesco Redi, showed in 1668 that meat protected from flies by being wrapped in muslin never produced worms, but meat that was left unwrapped developed "worms" (actually larvae) after flies laid eggs on it. Redi's experiments had been inspired by a passage in the *Iliad* in which Achilles asks his mother to watch over the corpse of his friend so that flies would not settle upon it and "breed worms about his wounds, so that his body, now he is dead, will be disfigured and the flesh will rot." A century after Redi, the Italian biologist Lazzaro Spallazani demonstrated that sterilized broths remained free of life if protected from airborne microorganisms.

The idea of spontaneous generation was essentially abandoned in 1862 after a series of careful experiments by Louis Pasteur. He showed that sterilized material protected from contamination in meticulously cleaned glassware remained devoid of microorganisms and insects indefinitely. Extrapolation of Pasteur's results into the distant past led some scientists to search for more exotic means of starting life on Earth.

In 1865 H. E. Richter suggested that life may have started on Earth after the arrival of microorganisms from space. He proposed that rapidly rotating celestial bodies might shed solid debris containing microorganisms. Such microbe-laden rocks would wander through interstellar space until a few landed fortuitously on a planet with a hospitable environment. Richter pointed out that it was plausible that the microorganisms would survive frictional heating during atmospheric passage because some meteorites contained carbon compounds that had successfully transited the atmosphere without burning.

So impressed was the British physicist William Thomson (later Lord Kelvin) with Pasteur's results that he used his 1871 presidential address to the British Association for the Advancement of Science to declare: "Dead matter cannot become living without coming under the influence of matter previously alive. This seems to be as secure a teaching of science as the law of gravitation." He rejected Charles Darwin's suggestion that life on Earth began in "some warm little pond" and held that life could not ever have formed from nonliving matter. He proposed that "the beginning of vegetable life on Earth might have started with the fall of a seed-bearing meteorite."

The German physicist and physiologist Hermann von Helmholtz said in 1874 that because scientific efforts had failed to produce organisms from nonliving matter, it appeared correct "to raise the question whether life has ever arisen" or "whether it is not just as old as matter." The French botanist P. van Tieghem wrote in 1890 that if life "like the universe itself, is eternal," then it must have spread to Earth from outer space.

In 1907 the Swedish Nobel laureate chemist Svante Arrhenius suggested that terrestrial microorganisms were sometimes blown into the stratosphere by high winds or thermal currents. Once there, they were apt to be ejected from the planet by electrical forces and propelled into the outer reaches of the solar system (and on into interstellar space) by radiation pressure from the Sun. Arrhenius reasoned that if microorganisms could escape the confines of Earth, an analogous event could have occurred on another planet around a distant star. Microorganisms expelled from such a planet could have found their way to Earth billions of years ago, colonizing our previously sterile world.

In the 1966 book *Intelligent Life in the Universe*, Carl Sagan showed that the fate of particles (living or nonliving) expelled into interplanetary space depends on the ratio of two forces, that due to radiation pressure from the Sun to that due to the gravitational attraction of the Sun. The net force toward or away from the Sun depends on the size of the particle. Sagan found that spherical microorganisms between 0.4 and 1.2 micrometers in diameter would escape the solar system. This size is smaller than that of many terrestrial microorganisms but similar to that of bacterial spores, fungal spores, and many viruses.

Microorganisms in this size range expelled from Earth would be accelerated by solar radiation pressure. They would reach the orbit of Mars in a few weeks and the orbit of Jupiter in a few months. Beyond the solar system, collisions with interstellar dust would slow the microorganisms, and they would likely take hundreds of thousands of years to travel a few light-years. But because stars in the solar neighborhood are randomly distributed most microorganisms traveling a few light-years would not be in the vicinity of a star; in fact, the average time it would take a microorganism to reach a new solar system is several hundred million years.

The principal question about extraterrestrial microorganisms seeding Earth concerns their survivability for hundreds of millions of years in interstellar space under conditions of extremely low temperatures, high vacuum, and intense radiation. In fact, many species of bacteria seem capable of enduring harsh conditions; some become dormant and form spores during periods of environmental stress. In this state they are resistant to boiling water, freezing temperatures, many chemical poisons, and the lack of water and nutrients. Halophilic (salt-loving) microorganisms do not form spores but are still hardy enough to survive a two-week trip in Earth orbit. Some bacteria such as *Deinococcus radiodurans* R1 are highly resistant to ionizing radiation.

In 1969 *Apollo 12* astronauts retrieved the television camera aboard the *Surveyor 3* unmanned spacecraft that had landed on the Moon thirty-one months previously. Upon examination, NASA scientists found live specimens of *Streptococcus mitis* bacteria inside the camera that had survived low temperatures, large temperature fluctuations, the absence of water and the vacuum of

Figure 17.1. Comet Hale-Bopp, photographed on 2 April 1997. There are two tails of the comet visible: the lower tail is made of dust that escaped the comet nucleus; the upper (near vertical) tail is composed of ionized gas. (© O. Richard Norton, Science Graphics; used with permission)

space. Although experiments have not shown that bacteria can remain viable for hundreds of millions of years, Raúl Cano and Monica Borucki reported in 1995 the revival of bacterial spores extracted from the digestive tracks of fossilized bees entrapped in 25- to 40-million-year-old amber.

However, bacteria traveling inside spacecraft or covered by amber have an external protective shield. Sagan showed that even if the hypothetical unshielded microorganisms escaping Earth had the radiation sensitivity of the most radiation-resistant bacteria known, ultraviolet (UV) light from the Sun with wavelengths smaller than 3,000 angstroms would kill them within a day. Failing that, x-rays and solar protons would kill them within a few years. The same lethal effects would occur for interstellar microorganisms coming to Earth. Although they could conceivably drift into the solar system and infect the outer planets, if they came within 1 AU of the Sun they would be killed by ionizing radiation. Cosmic rays would also damage such unshielded microorganisms, although the low energy of most cosmic rays might take tens of millions of years to provide a lethal radiation dose. Nevertheless, this is shorter than the probable mean travel time between stars. It thus seems likely that unshielded microbes cannot survive interstellar travel.

Yet there is another possibility: a journey shielded by natural materials. An object that could conceivably make an interstellar trip is a comet (figure 17.1). As the Sun revolves around the center of the Galaxy about every 240 million years, other stars occasionally pass near or through the Oort Cloud of comets, gravitationally perturbing some comets into new orbits. Some may be ejected into interstellar space and eventually get captured by another star. If such a comet someday collides with a hospitable planet, hypothetical microorganisms shielded deep within the comet could be released and revitalized.

However, the probability that the early Earth was struck by a comet from another solar system is quite low. Few interstellar comets would have been captured by the Sun; fewer would have collided with Earth, and, of these, fewer still would be expected to have struck during the first few hundred million years of Earth history. Furthermore, in our own solar system, comets are thought to have accreted in the vicinity of the giant planets and then to have been gravitationally perturbed into the Oort Cloud. The region where the giant planets formed is outside the life zone where liquid water is stable, and it is less likely that microorganisms would exist there. A possible exception would be liquid water beneath the icy surface of an orbiting satellite that was warmed by tidal friction. Even so, hypothetical organisms residing in such a subsurface ocean are unlikely to be incorporated into comets.

Another model of panspermia involving comets has been advanced by astrophysicists Fred Hoyle and Chandra Wickramasinghe. They pointed out that huge (20 light-year diameter) molecular clouds in interstellar space contain water, ammonia (NH_3), sulfur dioxide (SO_2), carbon monoxide (CO), and molecular hydrogen (H_2), as well as a rich variety of organic molecules including methyl alcohol (CH_3OH), ethyl alcohol (CH_3CH_2OH), formic acid (HCOOH), formaldehyde (H_2CO), methamimine (H_2CHN), acetylene (C_2H_2), and hydrogen cyanide (HCN). These molecules may have formed by adsorption of atoms on the surfaces of silicate or graphite dust grains that acted as substrates upon which chemical reactions could occur. Because molecular clouds are thick, ultraviolet radiation is blocked from much of the cloud, allowing organic molecules to form and survive.

Hoyle and Wickramasinghe postulated that two of the observed molecules, formic acid and methamimine, could react to form the simplest amino acid, glycine (NH_2CH_2COOH). They speculated that formaldehyde molecules could link up in interstellar space to form polysaccharide ($C_6H_{10}O_5$)$_n$ ring structures such as cellulose and starch. They claimed that in the Orion Nebula, interstellar dust produces a light-absorption curve identical to cellulose, while in some other regions, the absorption curve is essentially indistinguishable from that of dried batches of the bacterium *Escherichia coli.*

They suggested that intermittent outbreaks of human diseases such as influenza, the common cold, and smallpox are due to the infall of virus particles residing within micrometeorites derived from comets. The highly efficient mechanisms possessed by some bacteria and viruses to repair radiation damage from UV light were postulated to have evolved in space. Hoyle and Wickramasinghe asserted that if bacteria and viruses had instead evolved these radiation-repair mechanisms on Earth billions of years ago, when the atmosphere was thin and the UV flux was higher, these mechanisms would not have been conserved during evolution in the absence of selective pressures.

Variations on the idea that organisms abound in space have appeared in several classic science fiction stories including Michael Crichton's 1969 novel *The Andromeda Strain*, and the 1956 B-movie *Invasion of the Body Snatchers*.

A related variant of panspermia via comet delivery was suggested by NASA scientist Chris McKay in 1996. A generalized version of this scenario posits a solar system containing an inhabited planet wandering into a giant molecular cloud. Gravitational tugs by the cloud modify the orbits of comets at the fringes of the solar system causing some to collide with the inhabited planet. Rocks laden with microorganisms are blasted off the planet; some are perturbed into the molecular cloud and remain there after the solar system travels beyond the cloud. Millions of years later, the cloud collapses and forms new stars. The rocks containing the dormant microbes accrete with rocky, dusty, and icy debris to form comets at the nebular outskirts of the nascent stars. Gravitational perturbations from neighboring nebulae bring some of these comets toward their central star, where high ambient nebular temperatures cause them to evaporate. The microbes are released into interplanetary space, where many are killed by radiation. Some survive, however, and drift down to a new habitable planet; those surviving atmospheric passage colonize the surface.

Another possible way in which a planet can be infected by life from space is "directed panspermia" — the purposeful introduction of life on a planet by intelligent aliens. A science-fiction story employing this theme is James Blish's 1957 novel *The Seedling Stars*, wherein space-faring human geneticists travel the Galaxy seeding various planets with genetically engineered organisms possessing human intelligence.

The idea that life on Earth was purposefully introduced is an old one, forming the core of many religions. In Genesis 1 God's word permits the Earth to bring forth grass, herbs, fruit trees, beasts, and cattle, and the waters to bring forth "moving creatures," fowl, and whales. Man and woman are created directly. In a separate creation narrative in Genesis 2, man is formed of the dust of the ground. God breathes into his nostrils the breath of life.

Belief in directed panspermia is also the foundation of the Raëlian Movement, a large international organization that maintains that extraterrestrials called the Elohim (a biblical Hebrew plural noun translated as "God") arrived on Earth in the distant past to conduct biological experiments on a lifeless world. They created single-celled organisms and later genetically engineered more complex life-forms. Elohim scientists and artists collaborated to make beautiful scented plants and extravagant plumage in tropical birds. Eventually they created human beings in their own image. The Raëlian Movement was started in the early 1970s by a French journalist named Claude Vorilhon, who took the name Raël after claiming to have been contacted by an extraterrestrial who gave him the scoop on how life really started on Earth.

A variant of directed panspermia was proposed by astronomer Thomas Gold in 1960. He speculated that life on Earth evolved from microorganisms inadvertently left on Earth along with the trash by ancient space-traveling aliens with no compunctions against littering.

In a 1973 *Icarus* paper, Nobel laureate Francis Crick and fellow biologist Leslie Orgel postulated that because all terrestrial organisms have the same genetic code, all could have been descended from a single extraterrestrial microorganism. They suggested that microorganisms may have been seeded on lifeless planets throughout the Galaxy by an unmanned alien spaceship. Because some second-generation metal-rich stars similar to the Sun may have formed billions of years before the Sun, it is conceivable that there are technological alien civilizations in the Galaxy that are billions of years more advanced than we are. If we can foresee that our own civilization may one day be capable of such a project, then it is not out of the question that it has already been accomplished by an ancient alien civilization.

In his 1981 book on panspermia, *Life Itself: Its Origin and Nature*, Crick stated that the best choice of organisms to be sent to infect other worlds are anaerobic bacteria (organisms that thrive in the absence of oxygen) because in the absence of life, new worlds (like the prebiotic early Earth) are likely to lack molecular oxygen. Such anaerobic bacteria as *Escherichia coli* are single prokaryote cells that lack a nucleus and permit only objects of molecular size to pass through their membranes. Upon being frozen alive, stored for long periods of time, and eventually thawed, most individual *Escherichia coli* bacteria survive and reproduce. Crick wrote that if *Escherichia coli* were stored carefully, they might be able to survive for tens of thousands of years, probably even a million years. This length of time is probably long enough for spaceships traveling at 0.01 c (one one-hundredth the speed of light) to reach widely separated prebiotic planets across the Galaxy.

The newly formed Earth was an inhospitable place for life — it was too hot, extremely depleted in water, and subjected to too many high-energy impact events. As Earth cooled, colliding comets brought in water and organic compounds such as methyl cyanide (CH_3CN) and hydrogen cyanide (both of which have been detected in comet spectra); by 3.8 billion years ago, the impact rate declined significantly. Crick pointed out that the earliest known fossil microorganisms appeared on Earth not too long after conditions became favorable for life to exist. (Although this observation is consistent with panspermia, most biologists instead take this to indicate that life arises spontaneously and relatively quickly on planets with hospitable environments.) Crick concluded that directed panspermia is a valid scientific theory but one that would be difficult to test. Nevertheless, a couple of limited tests have been made:

1. If an alien civilization was clever enough to have seeded Earth with a microorganism, then it is possible that the aliens would also have encoded a message in the nucleotide sequence of its genome. It is conceivable that a legible remnant of the message would remain in the contemporary microorganism after billions of years of evolution. The trick is to guess which microorganism is a likely one to harbor a message. In 1979 researchers Hiromitsu Yokoo and Tairu Oshima published a paper in *Icarus* proposing that bacteriophage ϕX174 (a virus that infects the *Escherichia coli* bacterium in the human colon) was a good place to look for an interstellar message. They sequenced the DNA of a protein in the phage, made a pictogram of it, but perceived no orderly arrangement.

2. In the late 1960s Crick proposed that intelligently directed panspermia would involve an extremely hardy spore such as the one produced by the bacterium *Bacillus subtilis*. The complete 4,214,810-nucleotide sequence of this bacterium was published in *Nature* in 1997, but the 151 collaborating authors did not report finding a hidden message.

A different type of directed panspermia might be the radio transmission of the complete genome of selected organisms to other parts of the Galaxy. The transmission might be aimed at selected main-sequence stars thought to be plausible candidates for having life-bearing planets, or targeted to particular solar systems known to have intelligent life. Transmissions might be made in the hope that the receivers of the information would duplicate the genome and create for themselves an example of life from the transmitter's home world.

An inherent uncertainty in such a project would be how the receivers would use the information. For example, if we someday broadcast the complete human genome into space, it is conceivable that the information could be received by a technologically sophisticated but morally bankrupt alien society. Humans could be duplicated and placed in zoos or used as slaves. Even if the society had no need for cheap manual labor, slaves could be used for entertainment, sex, "scientific" experiments, conscripts, or dessert. Human captives have been used for all of these purposes in historical human societies.

If we one day received complete information about the genome of an advanced extraterrestrial being and had the technology to duplicate the organism, would we treat the individual with due respect, given our imperfect knowledge of its social, cultural, physical, and psychological needs? A related theme was explored in the 1995 science fiction movie *Species*, wherein government scientists decoded an alien radio transmission about a unique sequence of DNA and instructions on how to combine it with our own. The scientists followed these instructions with predictably disastrous results.

The most probable way for life to be transferred between worlds is by meteorites. In 1834 the Swedish chemist Jöns J. Berzelius reported carbon in the Alais CI carbonaceous chondrite (figure 17.2). Other chemical analyses of carbonaceous chondrites by Friedrich Wöhler and Elijah Harris in the 1850s and 1860s revealed organic material that Harris erroneously equated with a "bituminous substance," that is, coal. Because coal was believed (correctly) by most contemporary scientists to have been derived from decayed vegetable matter, the reported presence of coal in meteorites led some to believe that these meteorites had originated on a planet with life.

Hermann von Helmholtz and William Thomson were impressed by the occurrence of organic compounds in meteorites and suggested that these meteorites probably harbored life. Helmholtz gave lectures explaining that life would continue to exist in the universe even if, in the distant future, it somehow disappeared from Earth. Thomson cited Pasteur's experiments that disproved the spontaneous generation of life. He desired to refute Darwin's 1859 theory of evolution by natural selection because of its implication that life arose spontaneously from lifeless matter. Thomson declared that "overpoweringly strong proofs of intelligent

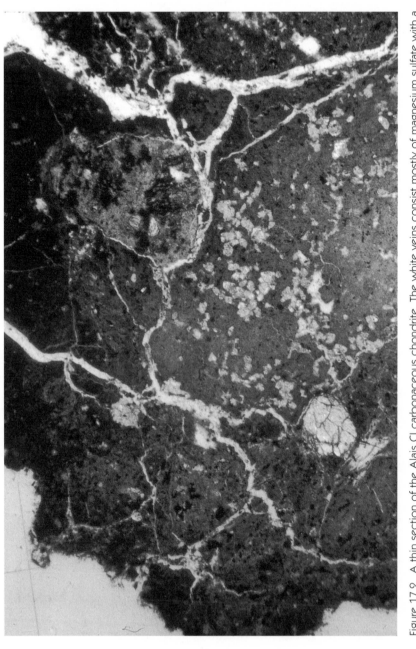

Figure 17.2. A thin section of the Alais CI carbonaceous chondrite. The white veins consist mostly of magnesium sulfate with a little calcium sulfate. The matrix of the meteorite consists mainly of hydrated silicate minerals. Field of view is 2 mm.

and benevolent design lie all round us . . . [and] that all living beings depend on one ever-acting Creator and Ruler." He went on to say that there was "no prospect of a process being found in any laboratory for making a living thing, whether the minutest germ of bacteriology or anything smaller or greater." Thomson reasoned that if scientists could not create life and if experiments had shown that life could not arise spontaneously, then life must have come to Earth from space.

The notion that microorganisms might survive in meteorites during the long time periods spent in interplanetary space appeared to be supported by experiments of the Hungarian botanist Gottlieb Haberlandt, who found that many seeds remained viable after long exposure to low temperatures in a vacuum.

The next step in deciding whether meteorites could have brought microorganisms to Earth in the distant past was to see if life could actually be detected in meteorites.

Otto Hahn, a lawyer and erstwhile geology student at the University of Tübingen in Germany, proposed that the common submillimeter-size spherules known as chondrules in the ordinary chondrite Knyahinya were plant fossils resembling algae and ferns. He also suggested that the Widmanstätten pattern in the Toluca iron meteorite was composed of fossilized plants (figure 17.3). (Modern meteorite researchers regard chondrules as having formed as silicate droplets during flash melting in the solar nebula and the Widmanstätten pattern in iron meteorites as the intergrowth of two distinct metallic iron-nickel minerals.) A zoologist named David Weinland collaborated with Hahn and published supporting findings. Weinland wrote that the meteorites contained corals that had turned to stone. He even named one species *Hahnia meteorica* in honor of his friend.

Several prominent scientists disparaged these results. Carl Vogt, Professor of Zoology at the University of Geneva, saw no similarity between the meteoritic inclusions and true corals. He said that "one does not know at what to be most astonished, the author's complete ignorance of the laws of evolution or the audacity with which he presents his views in terms worthy of the Oracle of

Figure 17.3. The Toluca iron meteorite showing the characteristic Widmanstätten pattern formed by the intergrowth of two iron-nickel minerals: kamacite and taenite. Maximum dimension of the slab is about 18 cm. (Photograph by Jeffrey J. Kurtzeman, Nininger Meteorite Award Poster. Courtesy of Carlton Moore, Arizona State University)

Delphi." These sentiments were echoed by J. Lawrence Smith, an experienced meteorite petrographer, who said that the common minerals in meteorites would not be confused with organic forms even by a beginner.

Nevertheless, the views of Hahn and Weinland were presented favorably in an 1882 issue of *Popular Science Monthly*. Since that time the notion that meteorites contain fossils or actual living organisms has been occasionally revived. The theme was explored in fiction in H. P. Lovecraft's 1927 horror story *The Colour out of Space*, wherein sinister space seeds within an iron meteorite poison the plants, animals, and people living on the farm where the meteorite fell.

In 1932 bacteriologist Charles B. Lipman from the University of California at Berkeley sterilized a suite of six ordinary chondrites, one achondrite, and one iron meteorite by immersing them in a 30 percent solution of hydrogen peroxide. He then grew bacterial cultures from their interiors over periods of one to two days. These cultures produced elongated and spheroidal bacterial cells closely resembling terrestrial forms. Lipman thought it reasonable that terrestriallike bacteria could have evolved on another planet. He declared himself the "pioneer of this subject," apparently forgetting that Louis Pasteur had unsuccessfully searched for microorganisms more than forty years earlier in the Orgueil CI carbonaceous chondrite.

Lipman's results were reported in the *New York Times* under the heading "Are Meteorites Alive?" However, the data were soon disputed by Yale bacteriologist Michael A. Farrell, who characterized them as "a flight of imagination" and by Harvey H. Nininger, the indefatigable meteorite field-worker, who suggested that the bacteria were terrestrial laboratory contaminants.

Sharat K. Roy, meteorite curator at the Field Museum of Natural History in Chicago, duplicated Lipman's experiments and found similar bacteria. He identified them as the common laboratory contaminants *Staphylococcus albus* and *Bacillus subtilis*. Roy questioned the logic of Lipman's conclusion: "If extraterrestrial bacteria could gain entrance into meteorites," Roy asked, "why could not terrestrial ones?" He went on to state that it was obvious that Lipman's "conclusion that he has discovered living visitors from the skies is merely his personal opinion and is not based on data that anyone searching for truth can or will ever accept."

After these reports were published in the mid-1930s, the idea that meteorites harbored life lay dormant until 1961, when Bartholomew S. Nagy of Fordham University and his coworkers analyzed Orgueil by mass spectrometry and found hydrocarbons similar to those produced by living terrestrial organisms. They claimed their results provided evidence for biologic activity. The *New York Times* quoted them as saying that "wherever this meteorite originated, something lived." Later that same year, George

Claus, a microbiologist at New York University Medical Center, and Nagy distinguished five varieties of "organized elements" in Orgueil and Ivuna (a similar meteorite) and concluded that they were possible remnants of organisms. Around this time, several micropaleontologists reported finding possible fossil forms in carbonaceous chondrites.

Other scientists pointed out that because hydrocarbons could be produced synthetically, their presence did not prove that meteorites harbored extraterrestrial life. Inorganic minerals could also assume ordered shapes; for example, ice can form snowflakes. Edward Anders and coworkers at the University of Chicago identified several of the varieties of organized elements that had been reported by Claus and Nagy as common meteoritic inorganic mineral grains. The others appeared to be common airborne contaminants such as ragweed pollen.

In 1982 Michael Engel and Nagy published a paper in *Nature* pointing out that amino acids in terrestrial organisms are all left-handed forms whereas amino acids produced nonbiologically are approximately half left-handed and half right-handed. Their analyses of amino acids in the Murchison CM carbonaceous chondrite showed more than 50 percent left-handed forms, suggesting biologic activity. However, because they could not rule out laboratory contamination, it seems likely that the biological activity they discerned was terrestrial. Claims by other workers in 1997 and 1998 that fossil bacteria had been discovered in Murchison and Orgueil were widely ignored by the scientific community.

An important question overlooked by these studies is whether the home worlds of meteorites are plausible abodes of life. The vast majority of meteorites come from asteroids—small rocky or metallic bodies that range in size from a few hundred meters to a few hundred kilometers. These bodies are too small to retain an atmosphere and too cold to maintain liquid water. They constitute an unlikely refuge for life.

More than a dozen meteorites are from the Moon (diameter 3,476 kilometers), a body far larger than any asteroid. Although

Nobel laureate chemist Harold C. Urey had suggested in 1959 that rocks blasted off the lunar surface by energetic impacts could fall to Earth as meteorites, the first lunar meteorite was not recognized until more than twenty years later. In a 1966 review paper, Urey restated the ideas that either the Moon was formed after escaping from Earth (in which case it could have been contaminated with terrestrial water) or it was captured by Earth (in which case it might contain some terrestrial material expelled during this violent event). He suggested that carbonaceous chondrites might come from the Moon and that they might contain primitive life-forms originating on Earth.

Although the bone-dry lunar rocks returned by the Apollo astronauts were not carbonaceous chondrites and did not contain living or fossil microorganisms, NASA made the *Apollo 11, 12, and 14* astronauts put on biological isolation garments before leaving their spacecraft, then quarantined them for three weeks to ensure that they did not harbor any hazardous lunar pathogens (figure 17.4).

The search for life in meteorites was revived in 1996 when NASA scientist David S. McKay and coworkers reported evidence of microscopic fossils of bacterialike organisms in Martian meteorite ALH84001. But, as discussed in the previous chapter, most researchers remained unconvinced.

Nevertheless, it is possible that primitive life once existed on Mars when conditions were more clement than today. It is even possible that Martian microorganisms gradually adapted to increasingly hostile conditions and still exist below the surface. Because Martian rocks have found their way to Earth as meteorites, it is conceivable that Martian life infected Earth billions of years ago and flourished after landing. If Earth was lifeless at the time of infection or if the Martian organisms were able to outcompete indigenous terrestrial organisms, then this variant of panspermia could account for life on Earth.

Before such scenarios can be considered plausible, many questions need to be answered: (1) Could impacts on Mars be capable of ejecting rocks large enough to permit effective shielding of

Figure 17.4. *Apollo 11* astronauts confined to the Mobile Quarantine Facility aboard the U.S.S. *Hornet* are greeted by President Richard Nixon. The astronauts are, *left to right*, Neil Armstrong, commander; Michael Collins, command module pilot; and Buzz Aldrin, lunar module pilot. (Courtesy of NASA)

internal microorganisms from galactic and solar cosmic rays? (2) Could Martian microorganisms survive the heat, shock pressure, and acceleration associated with launching rocks with velocities greater than the Martian escape velocity of 5 kilometers per second? (3) If the ejected rocks were large enough to shield the organisms from cosmic rays, could the bacteria survive the natural radioactivity of the rocks themselves? (4) Could the Martian microorganisms survive the intense cold and vacuum of interplanetary space? (5) Could the organisms survive spontaneous DNA decay?

In 2000 these questions were addressed in a paper published in *Icarus* by Curt Mileikowsky from the Royal Institute of Technology in Stockholm, Sweden, and his international team of coworkers. They pointed out that *Deinococcus radiodurans* R1 (the most radiation-resistant bacterium known on Earth) goes into a state of dormancy if there is a lack of food or water or if the temperature drops below the freezing point of water. The microbe *Bacillus subtilis* produces spores that are highly resistant to radiation and can withstand vacuum, lack of water, and low temperatures. Rocks large enough to shield these terrestrial microorganisms from cosmic rays have been blasted off the surface of Mars for billions of years. In 2001 Rachel Mastrapa and coworkers found that an appreciable fraction of bacteria survived extreme acceleration and change in acceleration (jerk) comparable to the motions expected from the launch of a rock off a planetary body. Mileikowsky and coworkers calculated that a small fraction of Martian rocks could make the trip to Earth in less than a million years. With such a short flight time, a significant fraction of these terrestrial bacteria (and, hence, of hypothetical Martian microbes) are likely to survive. They calculated that, over the past four billion years, there could have been five hundred million landings of Martian rocks on Earth under conditions that would have permitted *D. radiodurans* to have survived the trip. During the first five hundred million years of solar-system history, when collisions were much more frequent than today, there may have been five million landings on Earth of viable Martian microbes. Thus, if life was abundant on Mars in the distant past, it was probably brought to Earth.

However, because nearly all biologists believe that life originated on Earth (by the development of organic structures and functions from nonbiological precursors), it seems more likely that, if any interplanetary cross-fertilization occurred, life was brought from Earth to Mars. The presence of large ancient craters on the Moon indicates that, early in solar-system history, huge meteorites must also have struck Earth. Some of the ejected terrestrial rocks would have been launched at velocities greater than Earth's escape velocity of 11 kilometers per second. Some of these rocks would have reached Mars (about 2 percent of the number of Martian rocks reaching Earth). If terrestrial microorganisms within the ejected rocks could have survived the trip, they might have proliferated on Mars.

All terrestrial organisms, from aphids to apple trees, use the same DNA molecule. This indicates a common line of descent. It is possible that life beyond Earth uses molecules different from DNA. If this is the case and if Martian microorganisms are eventually discovered and found to contain DNA, this would support the idea that biological transfer has occurred between Earth and Mars.

Since interplanetary robotic travel began in the 1960s, terrestrial microorganisms trapped within spacecraft have reached the Moon, Mars, and Venus. Some have traveled on paths leading out of the solar system. Because inadvertent biological transportation has already occurred, interplanetary panspermia may be a reality.

The discovery of microorganisms on another planet would make headlines around the world, but when most laymen think of extraterrestrial life, they conjure images of ET. The discovery of intelligent aliens would connect Earth to the cosmos in a most dramatic fashion. However, the question of whether or not ET exists is not merely a matter for speculation. Plausibility arguments can be made, and these arguments form the subject of the next chapter.

XVIII

PAUCITY OF ALIENS

O

> Superior beings, when of late they saw
> A mortal man unfold all nature's law,
> Admired such wisdom in an earthly shape,
> And showed a NEWTON as we show an ape.
>   Could he, whose rules the rapid comet bind,
> Describe or fix one movement of his mind?
> Who saw its fires here rise and there descend,
> Explain his own beginning or his end?
> Alas, what wonder: man's superior part
> Unchecked may rise, and climb from art to art,
> But when his own great work is but begun,
> What reason weaves by passion is undone.
>
> —Alexander Pope, *An Essay on Man*

IT SEEMS that space aliens are everywhere. An issue of *UFO's Alien Encounters* magazine that I picked up in the mid-1990s featured stories about Nordic blondes from beyond, extraterrestrial gremlins, and flying saucers in Latin America. So many different kinds of alien space-travelers were described that, if we assume the veracity of the reports, Earth must be some sort of interstellar Grand Central Station.

Harvard psychologist John Mack, in his best-selling book *Abductions*, detailed his patients' fantastic stories of alien encounters. For example, one woman told Mack that she had been abducted more than one hundred times. Mack was impressed by the overall consistency of the reports, and it is clear that he found them credible. In the book *Close Encounters of the Fourth Kind*, journalist C.D.B. Bryan related reports of female abductees subjected to gynecological exams by extraterrestrials and of apparently open-minded individuals whose brains are periodically scanned by aliens. Bryan, too, was impressed by the consistency of the reports and concluded that since there is no proof that aliens don't exist, he would continue to keep an eye out for them.

In their 1992 booklet *Unusual Personal Experiences*, Budd Hopkins, David Jacobs, and Ron Westrum reported the results of a nationwide random survey of 5947 American adults. Two percent of the respondents met the authors' criteria as probable abductees by positively answering at least four out of five questions including whether they ever woke up paralyzed while sensing a nearby strange presence, saw unusual balls of light, had puzzling scars on their bodies, experienced a sense of mysterious flying, or had periods of an hour or more when they could not account for their whereabouts. The authors extrapolated the results of their survey and concluded that 3.7 million American adults have probably been abducted by aliens.

Skeptical Inquirer magazine reported that at the Bay Area UFO Expo in August 2000, Jacobs discussed spacecraft commanded by insectoid aliens. The ships were crammed with tables designed for human examinations. Hopkins warned of alien abductions of young children and infants. He cautioned that a sign of possible abduction is a baby showing fear when its mother wears dark glasses.

What is going on here? Are members of alien technological civilizations visiting Earth? Are they actually kidnapping our friends and neighbors? Are they really performing surgery without their patients' consent? Before these questions can be answered, a sober assessment is needed to evaluate the likelihood that there are any alien technological civilizations at all.

In the mid-1970s many scientists apparently thought that there were. Radio astronomer Frank Drake asserted, "At this minute with almost absolute certainty, radio waves sent forth by other intelligent civilizations are falling on the Earth." Papers published in *Icarus* during this period are rife with similar quasi-religious statements. John Ball believed that "extraterrestrial life may be almost ubiquitous" and concluded that our lack of interaction with aliens implies that Earth may have been set aside as a wilderness area or zoo.

Drake attempted to quantify the number (N) of technological civilizations in the Galaxy by writing a single equation with many unknown variables. A version of this equation is:

$$N = R^* \, f_{\text{suit}} \, f_{\text{plan}} \, n_{\text{suit}} \, f_{\text{orig}} \, f_{\text{cmplx}} \, f_{\text{intel}} \, f_{\text{comm}} \, L$$

Here, R^* is the mean rate of star formation in the galaxy, f_{suit} is the fraction of stars suitable for supporting life, f_{plan} is the fraction of those stars with a planetary system, n_{suit} is the number of planets per planetary system suitable for the origin and evolution of life, f_{orig} is the fraction of suitable planets where life actually arises, f_{cmplx} is the fraction of life-bearing planets where complex forms evolve, f_{intel} is the fraction of planets with complex life-forms where intelligent beings evolve who are capable of transforming their environment, f_{comm} is the fraction of planets where intelligent beings are capable of, and interested in, interstellar communication, and L is the mean lifetime of a technological civilization. Chemists Clifford Walters, Raymond Hoover, and R. K. Kotra introduced an additional term (C) into the Drake equation to "account for the effect of interstellar colonization."

Astronomers can estimate R^* fairly accurately. On average, approximately ten new stars form in our galaxy each year. Within the last few years, gravitational perturbations in the motions of nearby stars have led astronomers to infer the presence of Jupiter-size planets in orbit about them. In one case the shadow of an extrasolar planet was detected as a diminution of the brightness of its primary star as the planet passed in front it. Altogether more than seventy extrasolar planets had been discovered by summer 2001; more were expected to be announced in the following years. The spate of discoveries of these planets suggests that the term f_{plan} in the Drake equation is fairly high, say 0.5. The possible (albeit unlikely) existence of bacterialike micro-fossils in a meteorite blasted off the surface of Mars does not significantly augment the value of n_{suit}, which remains difficult to constrain. In most planetary systems the number is probably be-tween 0 and 3; the mean value is almost certainly much less than 1. The presumption that life arose on Earth within six hundred million years or so from the time of planet formation (at about the same time that intense meteoroid bombardment ceased) sug-gests that primitive life may arise fairly quickly on any suitable planet. It is even possible that life arose several times on the prim-itive Earth, only to be wiped out by gigantic impact events. It therefore seems reasonable to suppose that f_{orig} is fairly high, say 0.1. It is difficult to estimate f_{cmplx}, the fraction of suitable planets upon which complex life-forms evolve. It does not seem unreasonable that, given enough time, multicellular life might de-velop. Let's be generous and say that f_{cmplx} is also fairly high, perhaps also 0.1. Even Drake believes that f_{intel}, the fraction of those planets with complex life that go on to evolve intelligent life, is very small. It is not clear, however, what "very small" means: to Drake it might mean 0.1; to others it might mean 10^{-8}. The fraction of those planets that develop sophisticated radio technology and become communicative, f_{comm}, is also probably very small. Drake currently puts this value at about 0.25; others might choose a far smaller number. The mean lifetime of a tech-nological communicative civilization, L, is totally unconstrained. We have only one example, and we have been capable of sending radio, radar, and television signals to outer space for less than a century. Do some civilizations develop radio and nuclear weap-ons at about the same time and then blow themselves up? Do some civilizations last a billion years? If there are other civiliza-tions in the Galaxy, then the mean value of L will be dominated by the few very long-lived ones. Our profound ignorance of sev-eral terms in the Drake equation has led to estimates of N—the number of technological civilizations in the Galaxy—that range from 1 (ours) to 100 billion (which is of the same order as the number of stars in the Galaxy). This latter value would render the remarkable stories in *Alien Encounters* magazine slightly more plausible.

In their 1992 book *Is Anybody out There?* Drake and Dava Sobel estimated that there are at least 10,000 technological civilizations in the Galaxy; they predicted that we would detect radio signals from at least one of them before the year 2000. As of this writing, autumn 2001, no announcement of success has been made.

The purpose of this chapter is to show how recent research indicates that values of N significantly greater than 1 remain the province of science fiction writers, whether their work appears in *Analog* or *Icarus*.

For a planet to be suitable as an abode of life several constraints must be met. It must reside in the "life zone" around a single main-sequence star (or around a main-sequence star in a widely separated binary or multiple star system), where temperatures are moderate and liquid water is stable. Alternatively, there could be liquid water beneath the icy surfaces of outer planetary satellites that are tidally heated. Most double and multiple star systems (i.e., those systems in which the stars are not very widely separated) are ruled out because any planets they might have are likely to undergo wild orbital fluctuations causing extreme variations in temperature. Stars belonging to galactic or globular clusters are likely to encounter close passages of neighboring stars that might cause some of their planets to crash into their star or to be flung into interstellar space. Single stars that are many times more massive than the Sun will be short-lived and are likely to use up their hydrogen reserves in much less time than it would take complex life-forms to evolve.

The position of a star in the Galaxy is also a factor in determining whether complex life-forms are likely to emerge in an orbiting planet. The galactic center is the site of enormously energetic processes that might bathe planets there in lethal doses of radiation. The edges of the Galaxy contain many stars whose spectra indicate that they are poor in silicates and metals; they are unlikely to have Earthlike planets around them.

Caltech planetary scientist David Stevenson recently suggested that liquid water could exist (and life could conceivably develop) on Earth-size bodies composed of rock and ice that were ejected into interstellar space by gravitational scattering from Jupiter- or Saturn-size planets in nascent solar systems. If the Earth-size bodies were ejected while a molecular-hydrogen-rich nebula still existed, they might be able to retain much of the hydrogen as an atmosphere. These planets might retain sufficient amounts of their internally generated radioactive heat to allow temperatures to exceed the melting point of ice. However, even if such bodies exist (and Stevenson admits that they would be difficult to detect), the dearth of energy sources would probably severely limit the complexity of life-forms that could evolve. Such hypothetical bodies seem unlikely abodes of technological civilizations.

Another constraint on whether a planet is suitable for sustaining life is the mass of the planet, which must be sufficient for it to have a moderately thick (i.e., Earthlike) atmosphere (lunar-size bodies won't do) but not too great (comparable to that of the giant planets) or else nebular hydrogen and helium would be quantitatively retained. Liquid water could not exist at the surface of an airless body, but again, it could exist beneath the icy surface of tidally warmed satellites. The atmosphere of the planet must not trap too much stellar radiation or else, like Venus, it will produce a runaway greenhouse effect and cause extremely high surface temperatures, ridding the planet of liquid water.

Many of these constraints have long been recognized, but five additional requirements that drastically curtail the number of habitable planets in the Galaxy have recently come to light:

1. A Jupiter-size planet must occur in the outer part of the planetary system in order to deflect comets away from any habitable planets closer to the central star. Our solar system is surrounded by a diffuse collection of comets in the Oort Cloud, lying about 10,000 to 100,000 AU from the Sun. These comets probably accreted in the vicinity of the giant planets and were

flung out into the cloud by gravitational perturbations. This process is believed to have been quite inefficient; far more comets were probably ejected from the solar system than flung into the Oort Cloud. Thus, if Jupiter were absent from our planetary system (and replaced by a smaller body, the size of Uranus or Neptune, or by no body at all), the outer regions of the solar system would probably contain abundant cometary objects, some possibly exceeding the diameter of Pluto (about 2,400 kilometers). Such bodies could readily be perturbed into the inner solar system by the outer planets, causing catastrophic impacts on Earth every ten thousand to one hundred thousand years over the lifetime of the solar system. Such a scenario would make it difficult for terrestrial life to flourish. However, it is not clear how abundant Jupiter-size bodies are in the outer regions of other planetary systems. If they are rare, then any rocky planets in the inner regions of most planetary systems will probably be subject to frequent bombardment by cometary objects.

2. There must not be more than one Jupiter-size or super-Jupiter-size body in close proximity in the planetary system. Such objects could maintain stable orbits for millions of years or longer and then suffer chaotic gravitational interactions, forcing one another into highly eccentric orbits. This in turn would wreak havoc with the orbits of any rocky inner planets: some of these planets might move closer to the central star; others might be ejected from the system and plunged into the deep freeze of interstellar space. Only under the most fortuitous circumstances (e.g., wherein a rocky planet distant from the central star was perturbed into the life zone) would these interactions enhance the possibility of biological evolution.

3. A habitable planet in a system that contains additional planets must have a large Moon that stabilizes its obliquity (the tilt of its spin axis) in order to prevent extreme climatic fluctuations. Earth's axis is currently tilted about 23.3° from a line perpendicular to the plane of its orbit. This tilt gives us moderate seasonal variations in temperature and sunlight. Earth is not a perfect sphere; its spin causes the planet to bulge at the equator.

The Moon and the Sun pull on this bulge and cause the angle of the spin axis to vary over a period of 25,800 years. Recently, astronomer Jacques Laskar and colleagues from the Bureau of Longitudes in Paris ran computer simulations to determine what effect the absence of the Moon would have on Earth's spin axis. They found that gravitational tugs by the planets on Earth's bulge would cause Earth's obliquity to vary chaotically from about 0° to 50° over periods of a few million years. Over the long term the obliquity could even exceed 85°. This in turn would wreak ecological havoc by causing extreme climatic fluctuations. Epochs of enormous seasonal variations would be followed by periods with no seasons at all. Runaway effects might also come into play. For example, the climate could get cold enough to trigger extensive glaciation. Glaciers would reflect more of the incident sunlight into space and cause additional cooling. A "snowball Earth" situation could develop where ice covered the globe completely, as discussed in chapter 5. Tens of millions of years might pass before carbon dioxide emitted from undersea volcanoes caused the atmosphere to warm sufficiently to melt the glaciers.

What we really want to know is how many Earthlike planets in other planetary systems have moon-size satellites. The answer is probably very few. In our own solar system, the fact that no other terrestrial planet has a large moon suggests that the Earth-Moon system is unusual. The probable origin of the Moon as the product of a giant impact underscores the extreme rarity of such a system. It thus seems likely that "double planets" like the Earth-Moon system are very rare in the Galaxy.

This does not mean that life could not exist on a planet without a large moon. Terrestrial life is highly adaptable: algae survive on mountain peaks in Antarctica; bacteria thrive in scalding water near geysers; tube worms and giant clams flourish around volcanic vents far beneath the ocean surface. Nevertheless, drastic changes in climate every few million years would certainly constitute a formidable challenge to indigenous organisms, one likely to reduce significantly the number of planets with complex forms of life.

4. The existence of life in a planetary system probably depends on the bulk carbon to oxygen ratio (C/O) of the nebular material from which the planetary system formed. If the C/O ratio was high, much of the oxygen that was not incorporated into silicates would have combined with C to form CO and CO_2. Little oxygen would have been available to form water. The C/O ratios of stars are variable; those systems with high C/O ratios and consequently with little water are unlikely to harbor life.

The bulk carbon content of a planet is also an important constraint on whether it can support life. A planet that has little carbon cannot sustain a viable biosphere. A planet that has an overabundance of carbon will have a thick atmosphere rich in carbon dioxide and, like Venus, stifling surface temperatures. The amount of carbon is a function of the carbon content of the planetesimals that accreted to form the planet, the subsequent thermal history of the planet, and the number of late cometary and asteroidal collisions that brought carbon to the planet after it cooled. This latter number depends on the presence or absence of Jupiter-size planets in the outer regions of the planetary system.

5. As recently summarized by geologist Peter Ward and astronomer Don Brownlee, both from the University of Washington in Seattle, a habitable planet must have active plate tectonics. The benefits of drifting continents include diversity in climate, which in turn promotes biodiversity. A rich and diverse set of organisms is more likely to lead to complex life-forms. Plate tectonics are also responsible for recycling carbon, as limestone (rock made of calcium carbonate) on the ocean floor is drawn into the upper mantle, melted, and purged of carbon dioxide via nearby volcanoes. This process acts like a thermostat, moderating the temperature of the atmosphere and allowing liquid water to exist at the planet surface. Plate tectonics also seems to be indirectly responsible for Earth's magnetic field, which serves the important function of deflecting most cosmic rays away from the surface. Convection in the liquid outer iron core of Earth creates the magnetic field, but the convection cells would grind to a halt unless heat could be transported out of the core. Without plate tectonics

creating chains of volcanoes that serve as heat valves, the temperature of the outer core would become more uniform, convection would cease, and the magnetic field would disappear.

Even on those rare fortunate Earth-size worlds that orbit a single, stable, main-sequence dwarf star, that reside in a planetary system harboring a single Jupiter-size body in the outer regions, that have an appropriate amount of carbon, that have active global tectonics and are further blessed with oceans and a Moon, the chances that intelligent beings will evolve are still minute. This is because evolution is nondirectional. The diversity of life on Earth today is the sum of innumerable unpredictable events influenced by such forces as plate tectonics (which causes continents to join or split apart, thereby modifying climate, transforming ecological niches, and allowing populations to invade new territories or become isolated in old ones), asteroid and comet impact and, perhaps, extensive volcanism (which causes the extinction of some species and opens up opportunities for those which fortuitously survived), and chance occurrences such as freak storms that strand small populations on isolated islands. Evolutionary biologists are fond of pointing out that, if the clock were turned back and life started on Earth all over again, it is highly unlikely that any creatures resembling humans would again evolve. Indeed, it would be more probable to deal ordered suits from a shuffled deck of cards.

These contingencies are widely misunderstood by researchers outside the life sciences. For example, planetary scientist Thomas Heppenheimer, in his 1979 book *Toward Distant Suns*, estimated that there are 880,000 Earthlike planets in the Galaxy and concluded that each of them is likely to be the present or future home of an intelligent species.

To explore this fallacy, we need to look back at the evolution of life on Earth. Although life may have arisen as early as 3.9 billion years ago, the first organisms were simple prokaryotic cells (resembling modern bacteria and blue-green algae) lacking a membrane-bound nucleus. It probably took more than 1.5 billion years for the first complex eukaryotic cells to evolve. These cells have the capacity for sexual reproduction and contain a well-defined nucleus, helpful for storing and transmitting genetic information. Eukaryotes also contain small structures (organelles) such as mitochondria and chloroplasts that biologists believe were once independent prokaryote species that formed symbiotic relationships with the eukaryotes. The prokaryotes may have initially been eaten by the larger eukaryotes but not digested, or they may have been parasites that invaded the eukaryotes. In either case, the prokaryotes and eukaryotes formed a mutually beneficial biological alliance. The existence of eukaryotes begs a couple of questions: (1) *Was the formation of eukaryotes inevitable?* I do not believe that it was, but their evolution may have become increasingly probable with the passage of time and the evolution of prokaryote species that could become organelles. (2) *Why did it take so long?* Commensurate with the evolution of the prokaryote species that would form symbiotic relationships with the eukaryotes, the cells had to evolve a new architecture. This involved developing an outer cell wall and the ability to store the cell's DNA in a nucleus and compartmentalize other cell functions into the newly enclosed organelles. It seems possible that if Earth's clock were turned back and life started anew, the evolution of eukaryotes may have taken even longer or, just possibly, not happened at all.

About 570 million years ago, at the beginning of the geological period called the Cambrian, animals such as trilobites and brachiopods with hard calcium carbonate shells appeared and diversified. So many different types of fossils are found in Cambrian rocks (compared with the scarcity of fossils in older rocks) that this diversification of life is often called the "Cambrian explosion" (figure 18.1). But a survey of the organisms that flourished in this period would find that the members of our own phylum of chordates are few in number and not obviously destined for future evolutionary success as fish, amphibians, reptiles, and mammals. It seems quite plausible that a replay of Earth history would have featured no fish.

Mammals evolved more than 200 million years ago, about the

Figure 18.1. Artist's conception of the seafloor during the Cambrian period showing a sampling of the animals represented in the Burgess shale. The giant animal at top is *Anomalocaris*; the large stalks to the right are part of the sponge *Vauxia*. From left to right the animals along the sea floor are *Canadaspis, Sidneyia, Opabinia*, the trilobite *Olenoides*, and *Leanchoilia*. Two species of worms have burrowed into the sediment: *Louisella* on the left and *Ottoia* on the right. (Modified from S. C. Morris and H. B. Whittington (1985), Fossils of the Burgess shale: A national treasure in Yoho National Park, British Columbia, *Geol. Survey Canada Misc. Rpt.* 43, 1–31)

same time as dinosaurs (or just slightly later). Yet for the first 140 million years of their existence, mammals were small creatures relegated to the margins of a world dominated by dinosaurs. If it had not been for the impact of a 10-kilometer asteroid with the Yucatán peninsula 65 million years ago, dinosaurs would probably still be the dominant large vertebrates on land. As discussed in chapter 12, this impact event is thought to have caused worldwide cooling, months of darkness, acid rain, and global wildfires. Sixty-five to 70 percent of all living species perished. Most species of mammals survived because they were lucky, probably because they were small (and they were small because of the dinosaurs's success in occupying the ecological niches of larger land animals). Small animals are generally much more numerous than large animals; the greater numbers of individual animals gave mammal species a better chance of survival. The demise of the dinosaurs opened up fresh opportunities for mammals and allowed them to diversify. Without this impact there would have been no cows, cats, pigs, or people. If, however, Earth's clock were turned back and history started anew, the catastrophic impact would surely not have come 65 million years ago; it might not have come at all, or it might have come before mammals had evolved. There may even have been a second impact tens of millions of years later that wiped out most of the mammals. It is also possible that a larger projectile would have collided with the Earth 65 million years ago and caused the extinction of all multicellular organisms.

If Neanderthals had evolved and modern humans had not, technological civilizations would probably not have formed on our planet. Neanderthals seem to have lacked language, and their tools and artifacts show little change over their tenure.

Even after fully modern humans had evolved in Africa some 100,000 years ago, the eventual development of a technologically advanced civilization was not a foregone conclusion. Hunter-gatherers do not build radio telescopes (figure 18.2). Sources of readily extractable metal ore must be available for cultures to advance beyond Stone Age technologies. In addition, animals and plants that can be readily domesticated are required to permit some individuals to grow and store surplus food, thereby allowing others to specialize in technological, artistic, commercial, and military pursuits. Only a small fraction of wild mammal species have been domesticated because most lack one or more of the characteristics useful for domestication. The species should be nonterritorial and social (cats are an obvious exception). They should live in herds and transfer to younger generations their instinctive submissive behavior toward humans. They should be tolerant of intruders and able to breed in captivity. Sheep, pigs,

Figure 18.2. Antennas of the Very Large Array of radio telescopes near Socorro, New Mexico.

Figure 18.3. The major domesticated animals on Earth: a goat, a cow, a sheep, a pig, and a horse. Humans populating areas where these animals lived had a great advantage in establishing agriculture, building urban centers, and creating technological civilizations.

goats, cattle, and horses all fit the bill, and the occurrence of these animals in Eurasia aided Eurasian peoples in forming technologically advanced societies (figure 18.3). Conversely, the absence of such animals in Australia hindered the technological development of aboriginal Australians. It is possible that a replay of Earth history could have resulted in a world with no animals that could be domesticated. It follows that, on some planets, the resident intelligent beings may not be as lucky as the Eurasians and may never get around to inventing radios or rocket ships.

A final limiting factor on the number of technological civilizations in the Galaxy is the mean lifetime of these civilizations. Societies could succumb to disease, warfare, domestic political and economic instabilities, climatic changes, geological catastrophes (e.g., ice ages, earthquakes, and extensive volcanism), cosmic ca-

tastrophes (e.g., asteroid and comet impacts and nearby supernova explosions), and environmental crises (e.g., pollution, de-
spoliation of habitats, and depletion of natural resources). This
list is not exhaustive. There are far more than Four Horsemen
that could cause societal collapse.

It is even possible that some civilizations could lose interest in
interstellar communication and become galactic hermits. A variant of this theme was explored in 2000 by science fiction writer
Frederik Pohl in a short story called "The Brain Drain: Why Civilizations Fall Silent" that appeared in the science journal *Nature*.
Pohl suggested that denizens of technological societies download
their personalities into computer programs and lose interest in
the outside world.

Astronomers Ben Zuckerman and Michael Hart recently read-
dressed the conundrum (posed perhaps for the first time by En-
rico Fermi in the late 1940s) that if alien technological civilizations were abundant in the Galaxy, then they should be here
already. If astronomers were to discover around a nearby star an
Earthlike planet that had indications of harboring living organisms (e.g., its spectrum revealed an atmosphere containing water
vapor and ozone with traces of methane), biologists would be
clambering to send probes there as soon as it were technologically feasible. By analogy we can look at the timely exploration
of our own solar system. If we assume that some nearby intelligent aliens would be similarly motivated (which is likely if alien
civilizations are common), then the absence of evidence for contact suggests that alien technological civilizations are rare. One
might suppose that advanced alien civilizations may have no need
to visit us, perhaps having seen nascent technological societies
before (figure 18.4); however, if alien civilizations are abundant,
then at least some should be sufficiently motivated by curiosity to
pay us a visit.

In 2000 astronomer Ian Crawford of the University College
London pointed out in *Scientific American* that the first technological civilization capable of colonizing the Galaxy probably
could have done so before any competitors appeared on the
scene. He assumed that space ships could travel at 10 percent of

Figure 18.4. The Zoo Hypothesis maintains that Earth is a wildlife preserve watched by representatives of an advanced galactic civilization who refrain from visiting us and interfering in terrestrial affairs. In an inexact analogy, people watch a caged monkey hanging upside down at the Rio Grande Zoo in Albuquerque, New Mexico.

the speed of light, that colonies were spaced about 10 light-years apart, and that each colony would send out colonies of its own after a period of four hundred years. These figures indicate that the entire Galaxy could be colonized in about five million years. If the colonies waited an average of five thousand years before sending out colonies of their own, the Galaxy would be colonized in about fifty million years. Because billions of years have been available for technological civilizations to arise, one of them could have colonized the Galaxy and visited Earth hundreds of millions of years ago. *Chariots of the Gods* aside, there is no evidence that Earth has been visited by aliens in the past. If we wisely ignore the claims of UFO aficionados, then there is no evidence that aliens are visiting us at the present. We can therefore conclude that because aliens are not *here*, they may not be out *there*.

Another argument for the paucity of alien civilizations arises from probability theory, as explained by mathematician Amir D.

Aczel in his 1998 book *Probability 1*. If you throw a dart at a dartboard and are skilled enough to hit the board but are otherwise unable to direct the dart, it is more likely that the dart will strike a wider section of the board than a narrower one. If you were magically to become a piece of the dartboard and were randomly placed on it, it is more likely that you would end up in one of these wider sections. If you were to wake up tomorrow morning and discover that you had become sentient dartboard material, you would not have to look around at the rest of the board; chances are that you would be residing in one of the wider sections.

Likewise, if there is a broad, random distribution in the longevities of civilizations in the Galaxy, then chances are that you are in a comparatively long-lived one. In other words, if we assume that you have been randomly placed among the civilizations in the Galaxy, there is a greater chance of finding yourself in a long-lived civilization than in one that lasts only a short while. If we make the reasonable assumption that, on average, long-lived civilizations are more technologically sophisticated than short-lived ones, then it is probable that you find yourself among the most technologically advanced civilizations in the Galaxy.

If this is the case, then the fact that we developed advanced radio communications only in the last century implies that most other civilizations in the Galaxy (if any there be) are incapable of communicating with us. As we turn our ears to the sky, we will probably be greeted only by silence.

The sixteenth-century Italian philosopher Giordano Bruno wrote, "Innumerable suns exist; innumerable earths revolve about these suns. . . . Living beings inhabit these worlds." Although Bruno was probably correct in believing in the plurality of planets, and possibly also in the abundance of living organisms, he stretched his analogy too far. The probable paucity of alien technological civilizations suggests that, for all practical purposes, we are alone.

The topic of the final chapter is a simple one — but what if we're not?

XIX

HUMAN RESPONSE TO FIRST CONTACT

O

should far this from mankind's unmysteries
All nothing knowing particle who's i

look up,into not something called the sky

but(wild with midnight's millionary is)
a seething fearfully infinitude
of gladly glorying immortalities;

illimitable each transcending proud

most mind's diminutive how deathly guess
   —e. e. cummings

ONE OF the most profound events in human history, foretold thousands of times in fictional and philosophical writings, radio plays, movies, and television shows, would be mankind's first contact with an alien civilization. No single event would make more people feel connected to the cosmos than this. There are three basic ways in which this contact could occur: a direct visit by space-faring aliens (as depicted, for example, in such American films as *The Day the Earth Stood Still*, *Independence Day*, and *Star Trek: First Contact*), the discovery of an automated robotic probe (possibly akin to the one described in Arthur C. Clarke's *2001: A Space Odyssey*), and the detection of an electromagnetic signal (as in Carl Sagan's *Contact*). The signal could consist of rapid optical laser pulses or be a narrow-beam microwave beacon intentionally sent to us or intercepted fortuitously. It could also be unintentional radio leakage from domestic extraterrestrial broadcasts, perhaps an alien version of *The Beverly Hillbillies*. There are plausible variations of the main contact scenarios: for example, we could receive radio transmissions from a robot probe orbiting the Sun at the fringes of the solar system. It is conceivable that such a probe would be a conscious machine that initiated contact on its own volition.

We are at present technologically incapable of contacting alien civilizations by manned spacecraft or robotic probes; thus, if *we* are contacted by such direct means, it would indicate that the alien technology was significantly more advanced than our own. Likewise, because we have been capable of sending radio transmissions across the Galaxy for only a few decades, if we receive an alien signal from a star more than, say, 100 light-years away, it would mean that the alien broadcasters were technologically more advanced than we were when the signal was transmitted. Given the billions of years available for life to have evolved in the Galaxy, it is extremely improbable that civilizations on different planets would reach the same technological level at the same time. If there are relatively few technological civilizations in the Galaxy, as seems likely from our discussion in chapter 18, these societies would tend to be widely separated both in space and in technological sophistication. Because we could not detect a civilization significantly less technologically advanced than our own (they would lack radio), then, odds are, any signal we do receive would have been sent by a technologically more advanced society. Unintentional leakage from domestic alien broadcasts would probably be of low power and difficult to detect. Thus, if we ever acquire an alien radio signal, it is likely to be an intentional beacon.

It is easy for us to understand that an alien civilization interested in scientific exploration would listen for an artificial radio signal, but a problem arises if everyone is listening and no one is transmitting. This was the case for us from 1960 to 1974. In 1960 radio astronomer Frank Drake initiated Project Ozma, using the 27-meter antenna of the National Radio Astronomy Observatory in Green Bank, West Virginia, to search for radio signals from two nearby stars, Tau Ceti and Epsilon Eridani. In 1974 Drake used the newly upgraded 1,000-foot (305-meter) radio telescope in Arecibo, Puerto Rico, to transmit a message to the globular cluster M13 in the constellation Hercules. The message was a rather busy pictogram representing the binary numbers 1 through 10, illustrations of amino acids and the DNA double helix, a representation of a human figure, and schematic diagrams of the solar system and the Arecibo radio telescope itself (figure 19.1). The narrow band of the message gave it an effective radiated power of 20 trillion watts, making Earth brighter than the Sun at this particular wavelength for the three minutes it took to transmit the message.

Drake transmitted the message as a ceremonial gesture, provided mainly as entertainment for the 250 guests and speakers assembled in Arecibo for the telescope's rededication. The transmission was not a sustained effort. The chances of the message being received are slight: it would require an antenna in M13 (figure 19.2) to be scanning the sky in the Sun's direction at this particular wavelength twenty-four thousand years from now during the particular three-minute interval that the signal would be detectable.

In order for extraterrestrial technological civilizations to entertain reasonable prospects of making themselves known via radio, they would have to dedicate significant resources to a sustained transmission effort. What might be the motivation for such an enterprise? Of the numerous possibilities, there are ten that stand out. Of course, actual aliens may have different motivations. If we ever receive a signal, we could ask.

Exploration. A technological society might hope to receive a

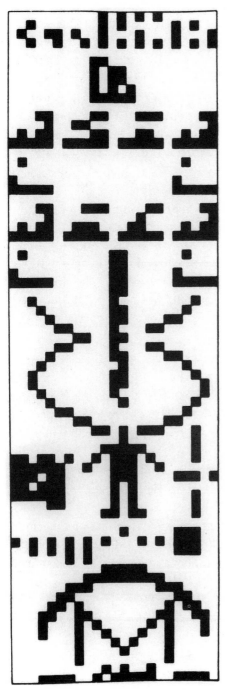

Figure 19.1. Pictogram sent by the Arecibo radio telescope in 1974. The message depicts the telescope itself (near bottom), the Sun and planets with Earth displaced upward, a man, a DNA molecule, atoms of hydrogen, carbon, nitrogen, oxygen, and phosphorus (elements necessary for life), and binary notations for the numbers 1 through 10.

Figure 19.2. The globular cluster M13 in the constellation Hercules, the intended target for the 1974 Arecibo message. Photographed by the 40-inch (1-m) Ritchey-Chrétien reflector. (Courtesy of U.S. Naval Observatory)

response by targeting signals to main-sequence stars that had been determined astrometrically to have planets. Perhaps radio leakage from a planet's domestic broadcasts was detected or ozone and methane was observed spectroscopically in a planet's atmosphere, indicating that life might be present. Perhaps an oxygen emission line was detected from a planet's auroral displays, indicating that photosynthesizing plant life was likely. It is also possible that the alien society was previously alerted to the presence of life on a planet by a robotic probe.

Hubris. A society may view itself as the glory of the universe and commit sufficient resources to broadcast this opinion. In this way it would be acting much like Ozymandias in Percy Bysshe Shelley's ironic poem. This ancient monarch had a gigantic stone statue of himself constructed with an immodest quote upon the pedestal: "My name is Ozymandias, king of kings: Look on my works, ye Mighty, and despair!"

Ostentation. Just as some Native American tribes in the Pacific Northwest engaged in potlatch rituals, giving away many valuable goods to display their wealth, some extraterrestrial societies might commit the resources to a dedicated, radio-transmission effort to impress themselves or rival groups of their wealth or status.

Evangelism. Technological societies, convinced that they had found the one true religion, might broadcast a message to save the souls of intelligent beings throughout the Galaxy. It could be an alien version of the Great Commission expressed in Matthew 28:19. "Go ye therefore, and teach all nations. . . ."

Entertainment. Extraterrestrials afflicted with ennui might find comedic value in transmitting unusual or humorous messages to the stars. The motivation would be similar to that of a bored cruise ship passenger who puts a joke in a bottle and casts it upon the sea.

Commerce. Entrepreneurial aliens may hope to establish a dialogue with another civilization in order to swap information on such topics as art, literature, music, history, and architecture.

They could use this information to produce and sell novelty items to their customers.

Altruism and paternalism. An ancient, advanced society may feel obligated to aid struggling newcomers to the galactic community with helpful advice on such topics as increasing agricultural productivity, avoiding nuclear war, eliminating pollution, and electing upstanding politicians.

Paranoia. A fearful alien civilization may feel compelled to broadcast threats throughout the Galaxy in order to intimidate unwanted potential visitors. Even though vast interstellar distances might render the threats idle, such transmissions may serve a psychological defense mechanism and soothe the aliens' fears. Extraterrestrials may also wish to represent themselves as repugnant monsters in hopes of deterring visits by curious space travelers.

Reproduction. An alien race with a sophisticated knowledge of biology and a determination to spread its seed throughout the cosmos might broadcast information about its genotype and instructions for creating breeding members of its species. It may adhere to a cultural or religious mandate to have descendants as numerous "as the stars of the heaven," as God promised Abraham in Genesis 22:17.

Pugnacity. An aggressive, mean-spirited alien society may derive pleasure from broadcasting threats to the cosmos, delighting at the prospect of instilling fear in potential listeners. It might provide detailed instructions on how to build powerful weapons, hoping that rival groups of recipients would use them on each other. It might even send deceitful friendly messages, hoping to identify and conquer unwary respondents within reach.

Irrespective of alien motivations, human reactions to the first contact with an extraterrestrial technological civilization would depend mainly on the nature of the contact, that is, whether it came by direct visit, robot probe, or radio transmission. If the contact was initiated by radio, our reactions would depend on a number of factors including whether the signal was recognized as artificial, if there was a period of social adjustment prior to official recognition of the signal as artificial, whether the signal could be deciphered, the content of the signal, and media coverage of the events.

If an anomalous radio transmission is received but its artificial character not generally recognized, there will be no social upheaval. Mitigating against our recognition of a real signal is the fact that narrow transmitting beams can be distorted by interstellar gas and magnetic fields. Radio waves can be refracted by gas and change direction; portions of the waves can shift and reinforce each other or cancel each other out. Despite these potential problems, alien radio transmissions may have already been detected. On 15 August 1977 the Ohio State radio telescope detected a transient, narrow-bandwidth extraterrestrial signal at a frequency of 21 centimeters; this is the same frequency emitted by the neutral hydrogen atom and has been singled out by astronomers as a likely frequency for interstellar radio communication. This particular signal fit so many of the criteria previously suggested as characteristic of an artificial transmission that radio astronomer Jerry Ehmann, the discoverer of the signal, wrote "Wow!" on the margin of the signal's computer printout. The only problem with the "Wow!" signal is that it has never been repeated. An artificial origin cannot be demonstrated, and it may well have been caused by some sort of radio interference. A few similarly strong, nonrepeating "Wow!"-like signals emanating from near the galactic plane were picked up by Harvard researchers. These signals are largely unknown outside SETI (search for extraterrestrial intelligence) circles and have provoked no public response.

It is conceivable that eventually many "Wow!"-like signals will have been detected and the news widely reported. Their possible extraterrestrial origin could be debated on the evening news, at congressional hearings, and in newspaper columns. If, in time, a consensus developed that the signals were indeed from an intelligent alien species, the public would be well prepared for the announcement. Little social upheaval would be expected.

Figure 19.3. Linear A script from a tablet from Crete found in the ruins of a small palace at a site known today as Hagia Triada. (After R. W. Hutchinson (1962), *Prehistoric Crete*, Middlesex, England: Penguin Books)

Even if a signal is recognized as being an alien transmission, it is not clear that it could be decoded. One might suppose that aliens would make an interstellar beacon easy to decipher, but this may not be the case. Furthermore, what is easy for them may not be easy for us. There are even human messages that have eluded decipherment for more than a century. The Linear A script (figure 19.3), first found in 1899 on pottery and clay tablets throughout Crete and other Aegean islands, was used by the Minoan civilization around 1500 B.C.E. Although we have abundant archaeological evidence of the Minoan culture, can empathize with the script writers as fellow human beings, and have even identified the Linear B script (which uses the same characters as Linear A) as an archaic form of Greek, Linear A has not been decoded. And this is not the only undeciphered human script: there are dozens of partially or completely unsolved ancient scripts including Mayan hieroglyphics (about 15 percent of these remain undeciphered), the Etruscan alphabet, and a puzzling Indus Valley script found on seals and copper plates. Extreme difficulties in translating numerous human scripts indicate that the decipherment of an alien radio signal is not guaranteed.

The content of a decoded alien message would greatly influence human reactions. A signal carrying only mathematical relationships or a friendly greeting would provoke less of an outcry than a blatant threat. Detection of radio leakage from domestic broadcasts of a distant peaceful society would also be likely to cause little anxiety. There are two imperfect precedents of human reactions to perceived alien detections.

In 1835 the *New York Sun* newspaper published a series of articles claiming that famed British astronomer Sir John Herschel had developed a new technique for boosting the magnification of his telescope. With these instruments Herschel was said to have observed diverse and benign life-forms on the Moon. The articles described a bucolic lunar landscape of oceans and beaches populated with bipedal beavers, bison, unicorns, miniature zebras, beautiful birds, and at least two species of bat-men. The latter were angellike creatures that flew like bats and walked like men. They were given the scientific-sounding Latin name *Vespertilio-homo* (literally, bat-man). The bat-men lived among blue sapphire temples and amethyst obelisks and spent their time in peaceful pursuits—flying, bathing, and sharing food with each other and with animals. The author of these stories, Richard A. Locke, may have been attempting satire rather than perpetrating a hoax, but the end result was commercial success for his newspaper. The *Sun* dramatically increased its circulation, becoming the most widely read newspaper in the world. Shortly afterward, a pamphlet containing reprints of the stories sold more than sixty thousand copies in a few days. The contemporary horror story writer Edgar Allan Poe appreciated the hoax and estimated that 90 percent of the *Sun*'s readers were taken in. The hoax was eventually exposed by the rival newspapers The *New York Herald* and *Journal of Commerce*. This perceived discovery of a non-threatening extraterrestrial civilization caused widespread interest but no public alarm.

In 1967 graduate student Jocelyn Bell detected periodic microwave pulses with the radio telescope at Cambridge University. Several sources were identified, and the possibility that these signals were artificial alien transmissions was seriously considered.

The radio sources were even initially called LGMs for Little Green Men. Numerous press reports raised the alien communication scenario, but again the public perceived no threat and the social fabric remained intact. The signals were soon concluded to be of natural origin and nicknamed pulsars, short for "pulsating stars." Astronomer Thomas Gold interpreted the signals as emanating from charged particles trapped within the enormous magnetic fields of rapidly rotating neutron stars. The discovery of pulsars won Bell's adviser, Anthony Hewish, who had had the telescope built to look for rapidly varying radio signals, a share of the 1974 Nobel prize in physics. Bell, herself, was not a corecipient.

Although there are no historical examples of a large number of people believing that scientists had detected a threatening extraterrestrial radio signal, there is the classic case of people believing that Earth had been invaded by evil aliens. On 30 October 1938, Halloween Eve, Orson Welles's *Mercury Theater on the Air* presented a radio play on CBS to a national audience based on H. G. Wells's novel *The War of the Worlds*. The play unfolded as a continuous news story with "reporters" describing a Martian invasion of New Jersey. "Enemy tripod machines" were mentioned along with deadly heat-rays and poisonous black smoke. Many people who had tuned in late and missed the initial credits accepted the reports as genuine. A few people panicked; some ran outside, barricaded themselves in their rooms, or jumped into their cars and sped away. An updated broadcast of the radio play in Chile in 1944 caused riots in Valparaiso. An angry mob burned down a radio station in Ecuador in 1949 after a rebroadcast of the play triggered widespread alarm. Six people died.

Although belief in the proximity of the Martians no doubt exacerbated listeners' anxieties, we can infer from these reactions that a threatening alien radio transmission would probably create a climate of fear. (Actually, the only reported victim of Martian attack is an Egyptian dog that died on 28 June 1911 when it was struck by one of about forty stones during the fall of the Martian meteorite Nakhla. However, the story may be apocryphal.)

Media coverage would also influence public response to the receipt of an extraterrestrial communication. Extensive, open, factual reporting of the event by professional journalists would go far in soothing people's anxieties. Tabloid sensationalism with hints of government cover-ups would likely generate public alarm, irrespective of the actual content of the message. There is a precedent for this as well.

The Space Age began with the launch of *Sputnik 1* by the Soviet Union on 4 October 1957, an event that took the American public by surprise. Although many thought of the Soviet Union as a Cold War adversary, most believed it was technologically backward. Being second in the space race left many Americans fearful. Rumors began circulating that the Soviets would soon put nuclear missiles on the Moon or in Earth orbit. Wishing to put rumors and speculations to rest, the directors of the Smithsonian Astrophysical Observatory tracked *Sputnik* and issued press releases on their findings. Reporters began checking with the observatory before printing unsubstantiated stories. This openness went far in calming American fears of the Soviet threat. It serves as an example that openness and accuracy would best serve society's interests if an alien radio signal should ever be detected.

Openness is desirable whether or not the signal is picked up by radio astronomers in the West or in a country with a long tradition of secrecy. The latter possibility, depicted in the 1968 novel *The Cassiopeia Affair* by Chloe Zerwick and Harrison Brown, suggests that there might be reasons for governments to want to suppress the news of first contact. In a 1990 article in *Acta Astronautica*, author Allen Tough listed four such reasons: (1) Governments may be unsure that the signal is genuine and suppress the news of contact in order to avoid ridicule if the signal is later shown to be of natural origin or a hoax. (2) Governments may fear riots and widespread panic more severe than that which resulted from the radio broadcasts of *The War of the Worlds*. (3) They may fear unraveling of social institutions, disruption of the economy, abandonment of established religions, and destruction of the political system. (4) Finally, governments may believe that the alien message may hold vital technological information that, if suppressed, would give them military or industrial advantages over their political or economic rivals.

Regardless of any governmental attempt to suppress the news of alien contact, it is unlikely to be kept secret for long, at least in the West. For example, a 1970 report from the Defense Science Board Task Force on Secrecy stated with regard to "vital information, [that] one should not rely on effective secrecy for more than one year." Researchers telephone, fax, and e-mail their colleagues frequently, gossiping about their latest findings. They brief one another at scientific meetings, in both formal and informal sessions. Radio astronomers are no more reticent to discuss exciting new results than anyone else. News of an unambiguous artificial alien signal would probably be leaked to the public in short order.

A cottage industry of speculation has developed about the human response to first contact. There are five main schools of thought.

The *Cure for Cancer Camp*. The optimists' school was founded by astronomers Carl Sagan and Frank Drake. In a 1975 article in *Scientific American*, they asserted that receipt of alien communication would "enrich mankind beyond measure." In this and other publications, they suggested that information derived from alien radio transmissions might allow us to cure cancer, achieve world peace, solve scientific puzzles, develop new art forms, and gain advanced technology. They dismissed the notion that shortcuts to advanced technology might carry such unintended negative consequences as displaced workers, overpopulation, psychological stress, and social unrest (if people came to believe that their governments were powerless or irrelevant). Nevertheless, the possibility of social upheaval is real: one need only examine the massive cultural disruptions of technologically primitive peoples such as the Yąnomamö of the Amazon Basin and Papuans in the interior of western New Guinea who came into contact with Western explorers in recent times.

The *End of Human Motivation Camp*. The pessimists' school was given birth by the fear of negative consequences of contact. In past decades the group claimed numerous adherents including science writer Ian Ridpath and two Nobel laureates—Harvard biologist George Wald and British astronomer royal Sir Martin Ryle. Wald believed that receiving information from advanced extraterrestrials would be like attaching ourselves to alien wisdom through an umbilical cord. He was terrified that such dependency could stifle human development and "fold the whole human enterprise—the arts, literature, science, the dignity, the worth [and] the meaning of man." However, most SETI enthusiasts find this faction much too pessimistic. For example, MIT physicist Philip Morrison argued that recognition and decipherment of alien communications and the subsequent cultural effects would likely constitute a slow process more akin to the development and spread of agriculture than the European discovery of America. Others pointed out that the receipt and decoding of information-rich alien transmissions would open up numerous opportunities for radio astronomers, cryptographers, anthropologists, biomedical researchers, industrial engineers, and perhaps even literary and film critics. "Well, Roger, it's the same old story of *skreejum* meets *wompbag*. Utterly predictable."

The *Downfall of Traditional Religion Camp*. Those who respond in terms of Christian beliefs comprise by far the oldest school. The stage was set for its debut by the fifteenth-century French theologian William Vorilong, who concluded that beings on other worlds would not exist in sin, because they did not spring from Adam. He acknowledged that Christ could have saved extraterrestrials by dying on Earth but declared that "it would not be fitting for Him to go unto another world that He must die again." From this perspective it follows that if aliens are some day discovered and found to be sinful, then there must have been an extraterrestrial Fall. Therefore, in this context, either the aliens would be unsaved, would have been saved by Christ's death on Earth but would be unaware of this fact, or would have been saved by Christ dying on their home planet as well as on Earth. These themes were explored in the 1793 book *The Age of Reason,* by British-born American revolutionary author Thomas Paine. Paine concluded that if Christianity is true and aliens exist, then "the person who is irreverently called the Son of God, and sometimes God himself, would have nothing else to do than to travel from world to world, in an endless succession of death, with scarcely a momentary interval of life." Paine was a deist and believed that astronomical research had established that extraterrestrials probably existed. He hoped his readers would follow his reductio ad absurdum and reject Christianity.

There are several ways out of this theological dilemma for fundamentalist Christians. For non-Christians, there is no dilemma.

The first way out was offered by Timothy Dwight, president of Yale University from 1795 to 1817. He responded to Paine's attacks by asserting that although intelligent beings probably inhabited other planets, only mankind had fallen into sin and required salvation. This idea was echoed in 1890 by Ellen G. White, founder of the Seventh-Day Adventist Church. She wrote in *The Story of Patriarchs and Prophets* that it was the "marvel of the universe that Christ should humble himself to save fallen man" and that this was "a mystery which the sinless intelligences of other worlds desired to understand."

Another way to meet Paine's challenge was advanced by Oxford astronomer E. A. Milne in 1952. He suggested that Christ's death had also brought salvation to extraterrestrials and that, in the fullness of time, they would learn of this fact after radio communication had been established with Christians on Earth. Upholding this view is planetary scientist and Jesuit Brother Guy Consolmagno, who wrote in 2000 that the resurrection of Christ was "the definitive salvation event for the cosmos." He cited Paul's epistles to emphasize this point (e.g., Col. 1:20): "And, having made peace through the blood of his cross, by him to reconcile all things unto himself; by him, I say, whether they be things in earth, or things in heaven."

A third way out of Paine's dilemma was reported by scientist-author Paul Davies in his 1995 book *Are We Alone?* He noted that Jesuit Father George Coyne, director of the Vatican Observatory, had suggested that God was free to save extraterrestrials by means other than His physical sacrifice. This would allow redemption to occur on numerous worlds without the endless deicide that many Christians find repugnant.

Another way out for believing Christians is to assume that aliens do not exist. This was the approach taken by the British scientist William Whewell, master of Trinity College in Cambridge. His 1853 treatise *Of the Plurality of Worlds: An Essay*, published anonymously, presented arguments that none of the known planets seemed suitable abodes for life. Whewell said that because humans have inhabited Earth for only a small span of time, it seems reasonable that intelligence in the universe is confined to only a small region of space—Earth. The probable non-existence of aliens is also assumed by many modern Christian fundamentalists, who maintain that the spontaneous origin of life on Earth inferred by evolutionists should lead to a populated cosmos. For example, John D. Morris, president of the Institute of Creation Research in El Cajon, California, sent an e-mail message to me stating that "in all likelihood, intelligent, mortal, extraterrestrial beings do not exist on other planets. This position is supported by both scientific & implications of Biblical statements." The absence thus far of evidence for extraterrestrial beings is taken by Christian fundamentalists as support of the special creation of life on Earth.

The *Denial Camp*. Firsthand knowledge of alien radio transmissions would be confined to staff members of observatories. The public would be dependent on accurate information from these individuals and factual accounts in the news media. Inevitably, no matter how secure the evidence, some people would assert that the transmissions were a hoax. There are numerous precedents for this.

In 1976, just four years after the *Apollo 17* astronauts landed on the Moon, author Bill Kaysing published a provocative book,

We Never Went to the Moon: America's 30 Billion Dollar Swindle. Although at $14.50 for a recent paperback edition, one could argue that Kaysing's book is more of a swindle than the Apollo program, it is fascinating to see how seemingly incontrovertible evidence can be denied. While one might suppose that photographs and video images of astronauts on the Moon could be faked and that astronauts and NASA officials could lie, it would be impossible to fake the rocks. Apollo samples are distinct in their mineralogy, mineral chemistry, and texture from known terrestrial rocks and the vast majority of meteorites. The only samples they closely resemble are lunar rocks returned by the unmanned Soviet Luna probes and rare meteorites from Antarctica, Australia, and Africa that are inferred to be from the Moon.

Another example of denial of the obvious is that of several fringe groups that claim that the Holocaust never happened. Despite a deluge of eyewitness reports, testimony from concentration camp survivors, Nazi confessions, captured documents, photographs, and artifacts, Holocaust deniers maintain that there was no official Nazi policy to exterminate Jews. They claim that those Jews who died in the camps succumbed to disease, starvation, and overwork. Minor inconsistencies in survivor accounts are highlighted, gas chambers and crematoriums are claimed to have been used only for delousing, and the "missing" Jews are said to be living quietly in Siberia or other remote regions. Deniers twist accounts, pick and choose pieces of evidence that support their views, and ignore contrary evidence. Detection of an alien transmission is sure to elicit the same type of response from individuals and organizations intent on proving a government-sponsored hoax.

The Go Back to Sleep Camp. It seems likely that most people would be rather indifferent to the news that an extraterrestrial civilization exists. There are many more immediate pressing concerns such as drawing a paycheck, harvesting crops, having babies, and nursing sick relatives. Throughout history, most people have been largely unaffected by news of such momentous events as the European discovery of America, Heinrich Schliemann's excavation of Troy, the realization that the universe is expanding, the discovery of the structure of DNA, and the manned lunar landings. The announcement that a radio telescope, distant from one's home, has picked up an alien transmission might send some people to toy stores to buy ET dolls for their kids but will otherwise leave most engrossed in their daily routines. Furthermore, the public in the West has been so inundated with stories of UFO abductions and alien autopsies that the actual discovery of an artificial extraterrestrial signal would probably not come as a shock. In this vein, Northwestern University astronomer and UFO investigator J. Allen Hynek said at a lecture I attended in the mid-1970s that there would be no better way for aliens to get humans used to their existence than by buzzing around the skies now and then and playing cat and mouse with fighter planes.

If an alien signal was received and recognized as artificial, astronomers would be eager to find out as much as possible about its source. Different radio telescopes would simultaneously tune in to the transmission in order to pinpoint the position of the source in the sky. If a star was identified as the source, as seems likely, astronomers would observe it with optical telescopes on Earth and at a variety of wavelengths with telescopes in space. Over the ensuing years, the proper motion of the star would be measured and astronomers would use astrometric techniques to try to determine the masses and distances of accompanying planets. A transmission not adjusted to compensate for the originating planet's rotation could be used to determine the planet's rotation rate. Of course, the most important information would likely be contained in the signal. Radio astronomers, physicists, cryptographers, and mathematicians would work hard to decode the transmission.

If the signal was decoded and found to be threatening or if the message portrayed the alien society as disgusting or reprehensible, there would be no push to transmit a reply. There would be little chance for a meaningful dialogue if the alien civilization was many hundreds of light-years away; again, there would be no need to rush a response. If we did not reply to a signal, the aliens would not know that we had received it unless they already knew that we were here and routinely monitored leakage from our domestic news broadcasts.

A friendly alien message would prompt many groups to reply. This would be especially true if the alien civilization was on a planet less than 100 light-years away. There would be a debate as to who had the appropriate authority to respond — individual investigators, observatory directors, federal agencies, or international organizations. It might be difficult to formulate a unified response, and there might be independent organizations and rogue nations that would reply on their own.

What should the nature of the reply be? Although it would depend on the content of the alien transmission, we might look to our three previous attempts to send meaningful messages to alien beings. The first message, conceived by Carl Sagan and Frank Drake, was on a small metal plaque aboard the *Pioneer 10* and *11* spacecraft (figure 19.4). *Pioneer 10*, launched in 1972 to photograph Jupiter, was accelerated by Jupiter's gravity and became the first artificial object to be flung out of the solar system. The plaque has a drawing of a naked man and woman, a schematic diagram of the spacecraft, a representation of the solar system, a depiction of the hydrogen atom, and a map showing the distribution and pulse frequencies of pulsars seen from Earth at the time the Pioneer spacecraft were launched. Because pulsar frequencies decrease with time, there would be only one place and one time in the Galaxy (i.e., the Sun in 1972) where pulsars with these frequencies and at these relative locations would occur. In other words, aliens who understood the diagram should be able to pinpoint the Sun's position and determine when the spacecraft was launched.

Because encounters between technologically advanced societies and primitive ones have nearly always been detrimental to the technologically less sophisticated, some scientists believe that telling aliens where we are is foolish. In fact, UCLA physiologist

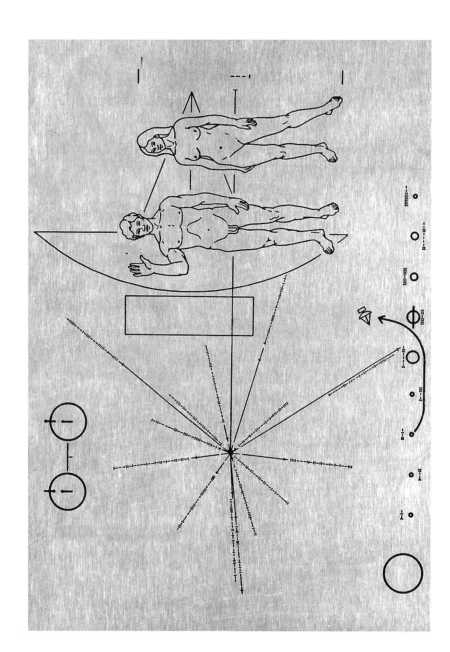

Jared Diamond stated that attaching these plaques to the Pioneer spacecraft was the "single stupidest, most dangerous act ever committed in human history!"

The second message meant for aliens was the pictogram transmitted in 1974 by Drake from the 1,000-foot Arecibo radio telescope toward the globular cluster M13 in Hercules. British astronomer royal Martin Ryle had concerns similar to those of Jared Diamond. After the Arecibo message was sent, Ryle urged the International Astronomical Union to ban all future transmissions.

The third interstellar message was a phonograph record aboard the Voyager spacecraft launched in 1977 to rendezvous with the outer planets and eventually leave the solar system. The record was produced by Sagan and Drake and contained such natural sounds as thunder, volcanoes, humpback whales, and crickets, selections of the world's music, greetings in sixty human languages, and an audio transcription of the electrical activity of a human brain.

Although it is doubtful that intelligent extraterrestrials will ever receive these messages, their transmission transforms *us* into a civilization attempting intragalactic communication. The Pioneer plaque and Voyager record will probably survive for tens of millions of years — longer than every other artifact so far produced by humankind. They seem destined to become the hoary emissaries of contemporary terrestrial civilization.

Figure 19.4. Plaque aboard the *Pioneer 10* and *11* spacecraft depicting a man and woman in front of a silhouette of the spacecraft. The "spider" at left shows the position of the Sun relative to the positions of 14 pulsars. At bottom are the Sun and planets showing that the spacecraft came from the third planet. At upper left is a representation of a neutral hydrogen atom; as the electron around the hydrogen atom jumps from a high-energy to a low-energy state, it emits radiation with a wavelength of 21 cm at a frequency of 1420 megahertz. This length and period are used on the plaque as units of length and time to show the distance to the pulsars and their pulsation periods.

It is hard to predict the ultimate consequences of the detection of alien messages. The only sure bet is that we would finally know that we were not alone and that other societies had survived their technological adolescence. On the other hand, if centuries pass and well-funded SETI programs turn up nothing, then we would know that, for all practical purposes, we were alone. Either finding would be fascinating.

EPILOGUE

O

The thirsty *Earth* soaks up the *Rain*,
And drinks, and gapes for drink again.
The *Plants* suck in the *Earth*, and are
With constant drinking fresh and fair.
The *Sea* it self, which one would think
Should have but little need of *Drink*,
Drinks ten thousand *Rivers* up,
So fill'd that they or'erflow the *Cup*.
The busie *Sun* (and one would guess
By's drunken fiery face no less)
Drinks up the *Sea*, and when h'as done,
The *Moon* and *Stars* drink up the *Sun*.

—Abraham Cowley, *Drinking*

IN THE latter half of the eighteenth century, Captain
James Cook undertook a series of historic voyages of exploration. He was sponsored by a powerful government agency (the
British Admiralty) and an influential private organization (The
Royal Society). Cook was charged with carrying out geographical, biological, and astronomical studies (figure E.1). One major
task was to search for a large, temperate southern continent. Although pack ice prevented him from discovering Antarctica,
Cook and his crew chalked up numerous successes. They charted
new islands, gave names to bays, rivers, capes, mountains, and

Figure E.1. A 1969 $1 coin from New Zealand commemorating the 200th anniversary of the voyage of Captain Cook to New Zealand.

promontories, collected botanical specimens, animal skins, birds, and insects (including many species previously unknown to Western science), made accurate drawings and paintings of native fish, birds, plants, and humans, and observed a transit of Venus across the face of the Sun.

Although the Venus observations of 3 June 1769 yielded unsatisfactory results due to the limitations of the available instruments, their purpose was a lofty one—to determine the dimensions of the solar system. If different observers at different locations on Earth had been able to time the transit of Venus accurately, the parallax angle between the observers could have been measured. Given the known distance between the observers, the Earth-Venus distance could have been directly determined. Since the relative distances of the planets were already known in Cook's time as fractions and multiples of the Earth-Sun distance of 1 AU, the absolute distance to Venus would have provided the scale of the solar system. Contemporary Royal Society astronomers could offer no better example of how Earth was an integral part of the solar system than by determining the absolute value of the astronomical unit.

Modern planetary scientists are following in the tradition of Captain Cook. Their work is sponsored by numerous government agencies (e.g., NASA; the National Science Foundation; the European Space Agency; the National Institute of Polar Research in Japan; CNES, the French space agency; the Vernadsky Institute of the Russian Academy of Sciences; the Max Planck Institutes in Germany) and influential private organizations (e.g., the Planetary Society; the W. M. Keck Foundation; the MacArthur Foundation; the Carnegie Institution; the J. S. Guggenheim Memorial Foundation). Like Cook, modern planetary scientists are engaged in geographical, biological, and astronomical studies (among others). They explore Earth from orbiting satellites, photograph the surfaces of other planets and moons, and give names to geographical features on these bodies. The recently founded Astrobiology Institutes in the United States fund studies of the earliest evidence of terrestrial life. The Planetary Society sponsors radio and optical searches for signals from any technologically advanced aliens who might be out there.

As of this writing, the *Pioneer 10* and the Voyager spacecraft are speeding toward the heliopause (the boundary between the solar wind and the interstellar medium), searching for the "edge" of the solar system. Once that boundary is crossed, the spacecraft should be surrounded by interstellar magnetic fields and particles unaffected by the solar wind.

In this book, we have followed scientists in their explorations of the solar system and in their search for life beyond Earth. We have seen how the solar system is interconnected and examined the roles played by impacts and gravitational perturbations. Humans indirectly owe their existence to the projectile that wiped out the dinosaurs, but human existence is threatened by other large bodies that may one day strike Earth. We need to catalog as many 50-meter or larger Earth-crossing asteroids as possible and keep an eye out for long-period comets.

But, as we have seen, asteroids are not merely potentially dangerous projectiles. Their study by earthbound telescopes, the Hubble Space Telescope, and spacecraft flybys and orbiters has enhanced our understanding of the early history of the solar system. The framework for this knowledge was built by laboratory-based analyses of meteorites — random chunks of asteroids that come to us for free.

The more we learn about our interplanetary backyard, the greater our appreciation for our seat on the porch. As planetary scientists continue their studies, we will find that (to paraphrase T. S. Eliot), "the end of all our exploring will be to arrive where we started and know [our] place for the first time."

GLOSSARY

O

accretion — The buildup of planets and planetesimals by low-relative-velocity collisions of smaller bodies in the solar nebula.

achondrite — Igneous stony meteorite that lacks chondrules.

acid rain — Rain that has an unusually high acidity.

agglomeration — Early accretion of chondritic constituents (chondrules, refractory inclusions, silicate matrix material) in the solar nebula to form small aggregations of chondritic matter.

albedo — The fraction of incident light reflected from a planetary surface.

algae — Eukaryotic single-celled, light-dependent microorganisms such as phytoplankton or multicellular seaweed.

alkali element — The elements from group 1 on the periodic table, i.e., lithium, sodium, potassium, rubidium, and cesium.

aluminosilicate — A mineral composed of oxides of aluminum and silicon.

amino acid — A small molecule containing $-NH_2$ at one end and $-COOH$ at the other that can link up with other amino acids to form a protein.

anaerobic bacteria — Prokaryotic microorganisms that can live without molecular oxygen.

angular momentum — The tendency of spinning or orbiting bodies to continue their motion due to inertia.

aphelion — The point in a body's orbit that is farthest from the Sun.

arcminute — A measure of apparent size equivalent to $\frac{1}{60}$ degree.

arcsecond — A measure of apparent size equivalent to $\frac{1}{60}$ arcminute or $\frac{1}{3600}$ degree.

arkose — A variety of sandstone containing abundant, poorly sorted grains of feldspar and quartz.

asteroid — A rocky, metallic, or icy body ranging in diameter from about 10 m to 1,000 km that is in heliocentric orbit.

asteroid belt — The region between Mars and Jupiter that contains the vast majority of asteroids.

asthenosphere — A portion of Earth's upper mantle that is partly molten and of lower viscosity than the immediately overlying lithospheric plates.

astronomical unit (AU) — The mean distance between Earth and the Sun, about 150 million km.

atomic number — A number equivalent to the number of protons in the nucleus of an atom.

aurora — High-altitude, multicolored, variable luminosity visible from the surface at night most often at high latitudes on Earth caused by the acceleration of charged particles (typically of solar origin) in Earth's magnetic field.

Australasian tektite — A tektite recovered from Australia or Asia.

background extinction — The typical rate of extinction of organisms throughout large periods of geologic time.

bacteriophage — A virus that infects bacteria.

baddeleyite — The mineral zirconium oxide (ZrO_2), often formed as a shock-induced breakdown product of zircon.

basalt — A dark, fine-grained igneous rock consisting mainly of the minerals plagioclase and pyroxene.

biofilm — A thin organic coating produced on surrounding mineral grains by moving bacteria.

black hole — A star that has collapsed in on itself to such an extent that the escape velocity at its surface exceeds the speed of light.

Bode's Law — An empirical mathematical relationship approximating the heliocentric distances of the planets Mercury through Uranus.

body-centered cubic — A crystal structure in which each atom has eight nearest neighbors.

bolide — A meteoric fireball, especially a meteoroid that breaks apart in the atmosphere.

brachiopod — A small marine invertebrate animal with a bivalve shell.

breccia — A rock that is a mechanical mixture of mineral and rock fragments; the proportions of these fragments and unbrecciated material can vary significantly.

brecciation — The process by which a coherent rock is broken into fragments and recemented.

calcareous — Containing abundant calcium carbonate.

caldera — A large volcanic crater formed by explosive eruption or collapse of a volcanic cone.

carbonaceous chondrite — A class of chondritic meteorites containing abundant silicate matrix material, relatively abundant refractory elements, and relatively small amounts of the isotope oxygen 17. Some groups contain several percent organic matter including amino acids. The CI carbonaceous chondrite group has elemental abundances very similar to that of the Sun's photosphere.

Cassini Division — A prominent 3,500-km-wide gap within Saturn's rings, located between the A and B rings. The gap contains several discrete ringlets.

catenae — Features consisting of aligned craters on the surface of a planet, asteroid, or moon, i.e., crater chains.

celestial mechanics — The science of the motion of celestial bodies governed by gravitational forces.

Cepheid variable — A class of variable star with a regular period that varies as a function of the star's luminosity.

chaotic terrain — Irregular, blocky surface depressions on Mars thought to be formed by subsidence due to the removal of subsurface water or ice.

chlorophyll — A green plant pigment found in photosynthetic organisms.

chloroplast — An organelle in plant cells that allows photosynthesis.

chondrite — The most abundant class of stony meteorite. Most groups contain chondrules, and all are similar in bulk composition to the Sun's photosphere. They are among the first solids to have formed in the solar nebula.

chondrite group — A collection of individual chondritic meteorites that have a narrow range in texture and bulk composition and are believed to have been derived from the same parent asteroid.

chondrule — A typically submillimeter-size crystalline and/or glassy quasi-spheroidal or ellipsoidal inclusion found in chondritic meteorites.

chordate — Animals belonging to the phylum Chordata, which includes all vertebrates.

chromosome — A linear strand of DNA and associated proteins in the nucleus of animal and plant cells that harbors the genes.

clast — A mineral or rock fragment embedded in another rock.

coesite — A silica mineral stable at high pressures that is often associated with terrestrial impact craters.

comet — A subplanet-size icy and dusty body that formed in the outer solar system and orbits the Sun, typically in highly elliptical or parabolic orbits.

condensate — In the solar nebula, a mineral phase that precipitates directly from a cooling vapor.

convection — The transfer of heat energy in a fluid by moving currents.

core — The central region of a body. In Earth and differentiated asteroids, the cores are rich in metallic iron-nickel.

corona — The outermost atmosphere of the Sun; it is very hot and extends far from the solar photosphere.

cosmic rays — Highly energetic atomic nuclei in outer space that bombard Earth from all directions.

cosmic-ray exposure (CRE) age — The period of time in which an object existed as a meter-size body in interplanetary space and was thus subject to irradiation by cosmic rays.

cosmochemistry — The study of the chemistry of planetary materials.

crater — A circular depression in the ground formed by a volcanic explosion or a meteorite impact.

Cretaceous-Tertiary boundary — The 65-million-year-old boundary between rocks deposited in the Cretaceous period and the succeeding Tertiary period.

crust — The outermost layer of Earth. Continental crust consists predominantly of granite and related rocks; oceanic crust consists mostly of basalt.

crustacean — An aquatic animal with a segmented body, exoskeleton, and paired, jointed limbs such as a lobster, crab, or shrimp.

crystal settling — A process by which crystals forming in a magma sink toward the bottom of the magma chamber under the influence of gravity and are unavailable for further reaction with the liquid.

Curie point — The temperature above which a mineral cannot remain permanently magnetized.

declination — The angle on Earth between the magnetic pole and the geographic pole.

deferent — An ancient geometrical construction used in geocentric models of the positions of planets and the Sun. It is a large circle centered on or near Earth on which an orbiting planet's epicycle (little circle) travels.

differentiation — A process in which a homogeneous chondritic body containing metal and silicate melts and forms distinct layers of different densities. In most cases, a metallic core and a silicate-rich mantle and crust are formed.

dipolar field — The main component of Earth's gravitational field, which resembles that of a bar magnet with north polarity at one end and south polarity at the other.

double star — Two gravitationally bound stars.

Drake equation — A relationship created by radio astronomer Frank Drake that attempts to calculate the number of technological civilizations in the Galaxy as a product of numerous uncertain parameters including the rate of star formation, the fraction of planets suitable for life, the fraction of those with intelligent life, and the longevity of technological civilizations.

echinoderm — A radially symmetric marine invertebrate such as a starfish, sea urchin, or sea cucumber.

eclipse — The partial or complete obscuration of one body by a second body as the second body passes in front of the first.

ecliptic — The plane of Earth's orbit around the Sun.

Edgeworth-Kuiper Belt — A group of comets residing mainly near the ecliptic at distances of 34–45 AU from the Sun.

ejecta — Material ejected from a crater due either to volcanism or meteorite impact.

electron microprobe — Instrument in which a beam of electrons is focused on a sample to produce x-rays. The amount and energy of the x-rays indicate the chemical composition of the sample.

ellipse — A plane curve drawn so that the sum of the distances from any point on the curve to two fixed points within the curve (the foci) is constant. The shape resembles an oval and is the characteristic orbital path of planets and asteroids around the Sun.

Encke Division — A 360-km-wide gap in Saturn's A ring.

epicycle — An ancient geometrical construction used in geocentric models of the positions of planets and the Sun. It is a small circle carrying a planet that moves around a larger circle (the deferent) that is centered on or near Earth.

equatorial bulge — The excess diameter of a spinning body due to mass buildup at the equator.

equilibrated chondrite — A chondrite with minerals of uniform composition (e.g., all of the olivine grains have the same composition) due to diffusion during thermal metamorphism. Such chondrites would be petrologic types -4 to -6.

erratic boulder — A boulder deposited on the surface by a retreating glacier that typically differs from local bedrock.

escape velocity — The velocity that an object must have to escape the primary gravitational influence of a larger body.

eukaryote — Single-celled or multicelled organisms whose cells contain a nucleus.

exothermic reaction — A chemical reaction that liberates heat energy.

extinction — The permanent disappearance of a species after all individuals have died off.

face-centered cubic — A crystal structure in which each atom has 12 nearest neighbors.

fall — A meteorite that was seen to fall.

feldspar — A group of alumino-silicate minerals containing a solid solution of calcium, sodium, and potassium.

find — A recovered meteorite that was not seen to fall.

fireball — A very bright meteor.

fissure — A narrow crack through which lava can extrude.

foraminifera — Marine organisms typically possessing a calcareous shell with small holes through which appendages can protrude.

fossil — Remains, track, or outline of an organism preserved in rock after the original organic material is removed or destroyed.

fossil meteorite — The textural, mineralogical, or compositional remnant within a sedimentary rock of a meteorite that fell millions of years ago.

fractional crystallization — A crystallization process in which minerals crystallizing from a magma are isolated from contact with the liquid.

fulgurite — A tubular, glassy object produced by melting materials at Earth's surface during a lightning strike.

fusion — In chemistry, melting. In nuclear physics, a nuclear reaction in which nuclei join together to make more massive nuclei and simultaneously release energy.

gabbro — A coarse-grained igneous rock of basaltic composition that formed at depth and consists mainly of plagioclase and pyroxene.

galactic disk — The flat part of a spiral galaxy that contains the spiral arms and appreciable amounts of gas and dust. It is the site of star formation. Stars in the disk generally revolve around the center in circular orbits.

galactic nucleus — The slightly flattened, bulbous central region of the galaxy within which stars move in elliptical orbits.

galactic year — The period of time it takes the Sun to revolve once around the galactic center, about 240 million years.

galaxy — A huge, gravitationally bound aggregation of stars, dust, and gas, typically with masses ranging from that of a few million to a few trillion times the mass of the Sun.

gamma ray — A photon of very high energy.

gastropod — A mollusk such as a snail or slug having a single coiled shell or no shell at all, a muscular foot, and eyes and feelers on the head.

gene — A hereditary unit that occurs in a specific location on a chromosome and, in concert with other genes, determines specific characteristics of an organism.

genome — A complete set of those chromosomes present in germ (ova or sperm) cells with the associated genes.

geocentric — Centered around Earth.

geomagnetism — Earth's magnetic field.

geyser — A spring that discharges hot water and steam into the air after groundwater is heated through contact with magma.

glacier — A large concentration of ice that does not completely melt in the summer and moves downhill under its own weight.

globular cluster — A spheroidal collection of about 50,000 to 1 million old, metal-poor stars that orbits around the galactic nucleus.

granite — An igneous rock consisting mainly of the minerals quartz and alkali feldspar.

graywacke — A variety of sandstone with a clay-rich matrix and containing poorly sorted rock fragments and abundant feldspar.

greenhouse effect — A warming of the surface of a planet caused by the absorption in the atmosphere of infrared radiation emitted by the planet after receiving a dose of solar radiation.

H I region — Region of neutral (un-ionized) hydrogen gas in interstellar space.

H II region — Region of ionized hydrogen in interstellar space, typically near very hot, highly luminous stars.

half-life — The period of time it takes half of a given number of atoms of a radioactive element to decay.

heliocentric — Centered around the Sun.

heliopause — The boundary between the solar wind and the interstellar medium.

herbivore — An animal that feeds mainly on plants.

Hertzsprung-Russell (H-R) diagram — A diagram of stellar luminosity against temperature, color, or spectral class that separates stars at different evolutionary stages.

highlands — The highly cratered, topographically high, ancient crust of the Moon; it is made mostly of a plagioclase-rich rock called anorthosite.

Hirayama family — A group of asteroids with similar orbital parameters formed by the collisional disruption of a single larger asteroid.

hydrocarbon — An organic compound composed of hydrogen and carbon arranged in rings or chains.

ice age — A period of time characterized by extensive glaciation.

ice sheet — Large continental glaciers, typically several kilometers thick.

immiscible — The property of liquids that are not mutually soluble such as oil and water or metallic and silicate melts.

impact basin — A large impact crater, typically many tens to hundreds of kilometers in diameter.

impact crater — A circular depression caused by compression and excavation of target material during the explosive impact of a projectile.

induction — The generation of electrical energy in a closed circuit due to a varying magnetic flux.

inertia — The tendency of an object to maintain its motion.

interplanetary dust particle (IDP) — Small dust particle in interplanetary space, a micrometeoroid.

interstellar dust — Dust grains in interstellar space.

ion — A positively or negatively charged atom.

ion microprobe — An instrument in which a focused beam of ions ionizes atoms in a sample and ejects them for analysis with a mass spectrometer.

iron meteorite — A meteorite consisting mainly of metallic iron-nickel minerals.

isotopes — Two or more atoms of the same chemical element that differ in the number of neutrons in their nucleus.

Ivory Coast tektite — A tektite recovered in or near the Ivory Coast in Africa.

kamacite — An iron-nickel mineral with less than about 8 percent nickel by weight that has a body-centered cubic structure.

Kirkwood gap — Paucity of main-belt asteroids with periods that are simple fractions of Jupiter's period due to resonance with Jupiter's gravitational influence.

Kuiper Belt — See Edgeworth-Kuiper Belt.

Lagrangian points — Mathematical points of gravitational stability in the orbital plane of a double star system, a planet, or a moon. Two of these points occur 60° ahead and 60° behind a planet in its orbit around the Sun; those associated with Jupiter mark the locations of the Trojan asteroids.

lava — Molten rock derived from a volcano.

lava tube — A roofed channel through which lava can flow.

layered tektite — A relatively large tektite made of multiple layers of glass; also called Muong Nong tektite.

lechatelierite — A form of silica glass.

Libyan Desert Glass — Massive, light-colored chunks of high-silica glass found in western Egypt near the Libyan border. The glass is most likely an impact-melt product similar to layered tektites.

lichen — Plants consisting of a fungus in close association with green or blue-green algae.

life zone — The range in distances from a central star where conditions are sufficiently clement to allow the possible existence of life.

limestone — A sedimentary rock composed of calcium carbonate.

lodestone — The iron oxide magnetite (Fe_3O_4).

loess — An unconsolidated, clay-rich sediment deposited by the wind.

magma — Molten volcanic rock beneath a planet's surface.

magnetic lines of force — The representation of a magnetic field in curved lines as indicated by the effects of the field on charged particles.

magnetic meridian — A great circle passing through Earth's magnetic poles.

magnetic polarity — Intrinsic separation of magnetic poles (regions of high magnetic intensity) with north at one end and south at the other.

magnetic reversal — A change in the polarity of Earth's magnetic field.

magnetite — A magnetic iron oxide mineral (Fe_3O_4).

magnetohydrodynamics — The study of electrically conducting fluids moving within electric and magnetic fields.

main-sequence stars — Stable stars that are converting hydrogen into helium in their cores. They form a diagonal swath on the H-R diagram.

mantle — The main, silicate-rich layer of Earth between the crust and the core.

maria — Dark lunar plains covered with flows of basalt.

mass extinction — The extinction of numerous species within a geologically short period of time.

mass spectrometer — An instrument that determines the chemical or isotopic composition of a substance by separating gaseous ions of the substance by their mass.

matrix — Fine-grained silicate-rich material in chondrites that surrounds chondrules, refractory inclusions, and other constituents.

mesosiderite — A stony-iron meteorite consisting of approximately 50 percent metallic iron-nickel and sulfide and 50 percent basaltic, gabbroic, and orthopyroxenitic silicates.

metamorphism — Change in the mineralogy and texture of a rock due to heat and pressure.

meteor — The light phenomenon produced by a solid object (a meteoroid) plunging through a planetary atmosphere and frictionally heating the surrounding air.

meteorite — A solid natural object reaching a planet's surface from interplanetary space.

meteorite shower — The simultaneous fall of numerous individual meteorites after the breakup of a single projectile in the atmosphere due to frictional stresses encountered during atmospheric passage.

meteoritics — The science involved in the study of meteorites and related materials.

meteoroid — A small solid object (larger than a dust particle and smaller than an asteroid) in interplanetary space.

meteor shower — The light phenomena produced by numerous small particles (meteoroids) traveling through a planet's atmosphere and heating the surrounding air. These typically occur at a particular time of year when Earth intersects low-density debris strung along the orbit of a comet.

methane — A gaseous hydrocarbon (CH_4).

microfossil — A fossil of a microorganism.

micrometeorite — A small meteorite, typically of millimeter size.

microorganism — A tiny, typically unicellular, organism.

microtektite — A tiny tektite, particularly those recovered in deep-sea drill cores.

microwave — Electromagnetic radiation with a wavelength in the approximate range of 1 mm to 1 m.

Milankovitch cycle — Periodic variations in Earth's orbital parameters that cause abnormal cooling and may be instrumental in the onset of ice ages.

mitochondria — An organelle found in eukaryotic cells that contains enzymes responsible for the conversion of food into usable energy.

moldavite — A tektite recovered within or near the Czech Republic.

molecular cloud — An interstellar gas cloud that is dense enough to allow the formation of molecules.

mollusk — A marine invertebrate having a soft unsegmented body, such as an edible shellfish or snail.

moraine — A deposit of till left at the margin of an ice sheet after the retreat of a glacier.

mullite — An aluminosilicate mineral ($Al_{4+x}Si_{2-2x}O_{10-x}$).

muon — A negatively charged subatomic particle with a mass 207 times that of an electron.

nebula — An immense, diffuse cloud of gas and dust from which a central star and surrounding planets and planetesimals condense and accrete.

North American tektite — A tektite recovered in North America, generally in Texas or Georgia.

nuclear fusion — The process in which atomic nuclei join together to make more massive nuclei and simultaneously release energy.

nucleotide — A compound consisting of a sugar linked with a phosphate group and an amino acid.

nucleus — In biology, an organelle containing chromosomes that is surrounded by a membrane within eukaryotic cells. In physics, the central part of an atom that contains the protons and neutrons.

obsidian — Silica-rich, dark volcanic glass.

Occam's razor — A philosophical principle, sometimes called the "law of parsimony," that is commonly interpreted to mean that the simplest of competing explanations is preferable.

occultation — The eclipse of one body by another such as when the Moon obscures the light of a distant star from the vantage point of an observer on Earth.

olivine — A mineral consisting of magnesium and iron silicate $(Mg,Fe)_2 SiO_4$.

Oort Cloud — Spherical cloud containing icy planetesimals extending out to about 50,000 AU from the Sun; this region is believed to be the source of some comets entering the inner solar system.

open cluster — A relatively small star cluster typically containing a few hundred to a few thousand stars. Typically, the stars are not densely packed together. Open clusters (also known as galactic clusters) occur mainly in the galactic disk.

orbit — The elliptical path of one body around another, typically the path of a small body around a larger body such as the Moon's path around Earth or Earth's path around the Sun.

orbital eccentricity — The deviation of an orbit from circularity; circles have eccentricities of 0.

orbital period — The length of time it takes an orbiting object to make one complete trip around its primary body. Earth's orbital period around the Sun is 1 year.

ordinary chondrite — The class of meteorites most common among witnessed falls. There are three principal groups: H chondrites (high total iron), L chondrites (low total iron), and LL chondrites (low total iron, low metallic iron).

organelle — A membrane-enclosed structure in a cell that performs a specific function.

outcrop — A portion of bedrock protruding above the soil.

ozone — A triatomic oxygen molecule (O_3).

paleomagnetism — The discipline of inferring Earth's ancient magnetic field and former continental positions through study of remanent magnetization in old rocks.

pallasite — A stony-iron meteorite consisting mainly of metallic iron-nickel and relatively coarse crystals of olivine.

Pangaea — The supercontinent that existed about 220 million years ago when plate tectonics had caused most of the major continental landmasses to collide.

panspermia — The hypothesis that life is spread from one planet to another.

parallax — The shift in the apparent position of an object when viewed from two different points.

parent body — The body from which a meteorite or meteoroid was derived prior to ejection.

parsec — The astronomical distance at which an object 1 AU across would subtend an angle of 1 second of arc. This distance is equivalent to 3.26 light-years, or 206,265 AU.

partial melting — The process whereby incompletely melted rocks form liquids differing in composition from the original composition of the rock.

partitioning — The chemical behavior of elements that causes them to prefer one mineral phase to another or to preferentially enter the solid or remain in the liquid during crystallization.

parturient montes, nascetur ridiculus mus — A line from the Roman poet Horace (65–8 B.C.E.) meaning "mountains will be in labor, and an absurd mouse will be born." In other words, there will be scant reward for all that work.

perihelion — The point in a body's orbit that is closest to the Sun.

perijove — The closest point to Jupiter in a body's orbit around Jupiter.

periodic comet — A comet with an orbital period of less than 200 years.

permafrost — Permanently frozen ice and soil found in cold regions.

petrology — The study of the composition, structure, and origin of rocks.

photosphere — Visible "surface" of the Sun.

photosynthesis — A metabolic process that uses light energy, carbon dioxide, and water typically to produce a simple sugar (glucose) and molecular oxygen.

phylum — A large taxonomic division of a kingdom of organisms, usually used in reference to the animal kingdom.

phytoplankton — Tiny aquatic plants that dwell near the surface of a body of water.

plagioclase — A feldspar mineral containing a solid solution of calcium and sodium.

planet — A substellar, moderate-size (more than about 1,000 km in diameter) body in orbit around a central star.

planetary nebula — An expanding, near-spherical cloud of hot gas ejected from and surrounding an old star.

planetesimal — A body ranging in size from a few meters to a few hundred kilometers that accreted early in solar-system history.

plankton — Small organisms dwelling near the surface of a body of water.

polycyclic aromatic hydrocarbons (PAH) — Organic compounds containing rings of six carbon atoms attached to hydrogen.

porosity — The volume percentage of a rock that consists of void space.

precession — The periodic shift in Earth's rotational axis due to external gravitational influences.

presolar grain — A mineral grain that formed before the solar system, fell into the early solar system, and was incorporated into chondritic material.

prokaryote — Unicellular microorganism lacking a nucleus.

pseudomorph — The occurrence of a mineral that has replaced a preexisting mineral and has retained the shape of the original phase.

pyroxene — A class of silicate minerals, most of which contain calcium and form a solid solution between iron and magnesium.

quasar — An ancient active galaxy with an intensely bright nucleus at enormous distances from Earth.

radiation pressure — The pressure exerted on tiny particles in interplanetary space by solar radiation due to absorption of light by the particles.

radioactivity — A property of some isotopes that causes their nuclei to undergo spontaneous disintegration resulting in the emission of subatomic particles.

radiometric age — The age of an object determined by the proportions of its original radioactive elements and their decay products.

radionuclide — A radioactive isotope.

rare-earth elements — Elements with atomic numbers 57 (lanthanum) to 71 (lutetium).

rare-earth pattern — The abundances of rare-earth elements relative to those in chondritic meteorites.

red giant — An old, highly luminous red star with a relatively cool surface that has burned most of the hydrogen in its core. Prominent examples include Betelgeuse in Orion, Antares in Scorpius, and Aldebaran in Taurus.

refractory element — An element that would condense at high temperatures from a gas; such elements include calcium, aluminum, titanium, iridium, and osmium.

refractory inclusion — An inclusion rich in refractory elements, particularly calcium, aluminum, and titanium in a chondrite. These are commonly known as calcium-aluminum inclusions or CAIs.

regolith — Unconsolidated and fragmented debris composed of rocks and minerals located at the surface of a body lacking an atmosphere.

remanent magnetism — Permanent magnetization acquired by rocks from Earth's magnetic field.

resonance — A dynamical relationship among bodies typically where a small body has an orbital period that is a simple fraction of a nearby larger body; the periodic gravitational tug of the large body causes the smaller body to leave its previous orbit.

Roche Limit — The mathematical distance from a massive body within which tidal forces would pull apart an orbiting object of zero internal strength (i.e., a fluid body).

RR Lyrae star — A variety of massive, pulsating variable star with a period of less than 1 day.

rubble pile — A small body consisting of jumbled rock fragments and void spaces.

sandstone — A sedimentary rock composed mostly of grains ranging from 0.0625 to 2 mm and consisting in most cases mainly of quartz, feldspar, and rock fragments.

semimajor axis — Half the length of the long axis of an elliptical orbit, equivalent to the mean distance of an orbiting object from its primary body.

SETI — Acronym for Search for Extraterrestrial Intelligence.

shepherding moon — Small moons within the ring systems of giant planets that gravitationally influence the distribution of ring particles.

shergottite — Igneous stony meteorite consisting mainly of plagioclase (or a shocked glass of plagioclase composition) and pyroxene and believed to come from Mars.

shield volcano — A broad, gently sloping volcano built up by the extrusion of low-viscosity lava.

siderophile element — A metallic element with a chemical affinity for metal. Such elements, which include nickel, gold, and iridium, mainly occur as metals rather than oxides.

Signor-Lipps effect — A statistical analysis in paleontology showing that the chances of finding a particular fossil in rocks close to a boundary decrease as the boundary is approached.

silica — The chemical compound of silicon dioxide (SiO_2).

silicate — A large class of minerals containing silicon and oxygen.

silt — Unconsolidated sediment consisting of small mineral particles intermediate in size between sand and clay.

sinuous rille — A narrow trench (particularly on the Moon) following a snakelike, meandering path.

solar flare — An abrupt, violent outburst of energy on the solar photosphere.

solar nebula — See **nebula**.

solar system — The Sun and set of objects orbiting around it including planets and their moons and rings, asteroids, comets, and meteoroids.

solar wind — The flow of ionized gas through the solar system originating in the Sun's corona.

solid solution — A series of minerals of the same structure but containing a mixture of different elements in widely varying proportions. A prime example is olivine, which varies continuously from the magnesium end-member (forsterite; Mg_2SiO_4) to the iron end-member (fayalite; Fe_2SiO_4).

spallation — The formation of new isotopes after irradiation of precursors in a meteoroid by cosmic rays.

spectral type — The color of a star including the presence of dark lines in its spectrum indicating the presence of certain chemical elements in the star's atmosphere.

spiral galaxy — A galaxy with prominent spiral arms that contain young stars and large amounts of gas and dust. Also present is a massive diffuse halo known as the galactic nucleus.

splash-form tektite — Tektites that have the shape they acquired as spinning molten objects. These include dumbbells, spheroids, teardrops, lenses, and buttons.

stishovite — A silica mineral formed from quartz in terrestrial craters during energetic, very high-pressure impact events.

stony-iron meteorite — An igneous meteorite that consists mainly of metallic iron-nickel and silicates. The two principal groups are pallasites and mesosiderites.

strata — Originally horizontal layers of rock.

stratosphere — An upper layer of Earth's atmosphere in which there are few clouds and only minor variations in temperature with altitude.

stratovolcano — A volcano consisting of solidified lava and fragmental rocks formed during volcanic explosions.

strewn field — A typically elliptical distribution of recovered meteorites after a meteorite shower.

striated rocks — Rocks with grooves, ridges, and scratches formed by glacial action.

sunspot — A temporary, relatively cool region on the Sun's photosphere associated with an intense magnetic field.

supernova — A highly energetic exploding star that produces heavy elements by neutron bombardment and ejects them into interstellar space.

taenite — An iron-nickel mineral with about 15–40 percent nickel by weight and having a face-centered cubic structure.

tektite — Glassy object produced by the melting of silica-rich sediments on Earth by the impact of an asteroid or comet.

Tharsis Ridge — A crustal bulge on Mars that is the site of about a dozen relatively young volcanoes and associated volcanic plains.

thermoluminescence (TL) — Emission of light caused by the heating of certain minerals.

thermoremanent magnetization — Permanent magnetization acquired by igneous rocks in the presence of a magnetic field as the rocks cool through the Curie point.

tidal heating — Frictional heating within a body caused by tidal stresses induced by differences in gravitational pull in different regions of the body.

till — An unconsolidated, poorly sorted sediment deposited by glaciers. Till contains rock and mineral fragments of all sizes, from clay-size particles to massive boulders.

trilobite — An extinct arthropod animal whose body structure consists of three lobes.

Trojan asteroids — Captured asteroids residing at the Lagrangian points in Jupiter's orbit, 60° ahead of and 60° behind Jupiter.

tsunami — A large destructive seismic sea wave caused by an earthquake.

T Tauri star — A pre-main-sequence star of approximately solar mass characterized by erratic outbursts, flares, and appreciable mass loss.

type-3 chondrite — An unequilibrated chondrite with a relatively unre-crystallized texture and containing mineral grains with heterogeneous compositions.

types -4 to -6 chondrites — Increasingly equilibrated, thermally metamorphosed chondrites.

unequilibrated chondrite — A chondrite with heterogeneous mineral compositions.

variable star — A star that is observed to vary in brightness either because of intrinsic instabilities or because it is periodically eclipsed by a neighboring object.

viscosity — The degree to which a liquid resists flow. High-viscosity liquids (e.g., "molasses in winter") resist flow to a great degree.

volatile element — An element that condenses from a gas at relatively low temperatures. Such elements include sodium, indium, gold, mercury, and the noble gases (e.g., helium, neon, argon, krypton, xenon).

volcanic cone — A small volcano, typically of conical shape.

volcanic dome — A bulbous or platy volcanic structure composed of stiff, pasty lava, typically rising from a fissure or inside a caldera.

Widmanstätten pattern — A texture in iron meteorites, enhanced by etching with a dilute solution of nitric acid, that shows the intergrowth of the two main iron-nickel minerals, kamacite and taenite.

x-rays — High-energy, short-wavelength photons.

zircon — The mineral zirconium silicate ($ZrSiO_4$).

zodiacal light — Sunlight scattered by interplanetary dust extending away from the Sun along the ecliptic and faintly visible in the eastern sky shortly before sunrise or in the western sky shortly after sunset.

ADDITIONAL READING

O

Chapter 1. A Brief History of the Solar System

Brush, S. G. 1990. Theories of the origin of the solar system, 1956–1985. *Reviews of Modern Physics* 62, 43–112.

Cohen, M. 1987. *In Darkness Born: The Story of Star Formation.* Cambridge: Cambridge University Press. 220 pp.

Crovisier, J., and T. Encrenaz. 2000. *Comet Science: The Study of Remnants from the Birth of the Solar System.* Cambridge: Cambridge University Press. 173 pp.

Dermott, S. F., ed. 1978. *The Origin of the Solar System.* New York: Wiley & Sons. 668 pp.

Gehrz, R. D., D. C. Black, and P. M. Solomon. 1984. The formation of stellar systems from interstellar molecular clouds. *Science* 224, 823–30.

Jones, B. W. 1999. *Discovering the Solar System.* Chichester: Wiley & Sons. 416 pp.

Taylor, S. R. 1992 *Solar System Evolution: A New Perspective.* Cambridge: Cambridge University Press. 307 pp.

———. 1998. *Destiny or Chance: Our Solar System and Its Place in the Cosmos.* Cambridge: Cambridge University Press. 229 pp.

Wood, J. A. 2000. *The Solar System.* 2nd edition, Upper Saddle River, N.J. Prentice-Hall. 209 pp.

Chapter 2. Where Are We? The Location of the Solar System

Impey, C. and W. K. Hartmann. 2000. *The Universe Revealed*, Pacific Grove, Calif.: Brooks/Cole. 636 pp.

Koestler, A. 1959. *The Sleepwalkers: A History of Man's Changing Vision of the Universe*. New York: Grosset and Dunlap. 624 pp.

Seeds, M. A. 1999. *Foundations of Astronomy*. Belmont, Calif.: Wadsworth Publishing Co. 640 pp.

Chapter 3. Heat Sources

Beatty, J. K., C. C. Petersen, and A. Chaikin, eds. 1998. *The New Solar System*. 4th edition. Cambridge: Cambridge University Press. 432 pp.

Hoyt, W. G. 1987. *Coon Mountain Controversies: Meteor Crater and the Development of Impact Theory*. Tucson: University of Arizona Press.

McSween, H. Y. 1999. *Meteorites and Their Parent Planets*. 2nd edition. Cambridge: Cambridge University Press. 310 pp.

Sonett, C. P., M. S. Giampapa, and M. S. Matthews, eds. 1991. *The Sun in Time*. Tucson: University of Arizona Press. 990 pp.

Taylor, S. R. 1992. *Solar System Evolution: A New Perspective*. Cambridge: Cambridge University Press. 307 pp.

———. 1998. *Destiny or Chance: Our Solar System and Its Place in the Cosmos*. Cambridge: Cambridge University Press. 229 pp.

Chapter 4. The Magnetic Earth

Brown, G. C., and A. E. Mussett. 1993. *The Inaccessible Earth: An Integrated View to its Structure and Composition*. 2nd edition. London: Chapman & Hall. 276 pp.

Cook, A. H. 1973. *Physics of the Earth and Planets*. New York: John Wiley & Sons. 316 pp.

Jacobs, J. A. 1975. *The Earth's Core*. London: Academic Press. 253 pp.

Khramov, A. N. 1987. *Paleomagnetology*. Berlin: Springer-Verlag. 308 pp.

Chapter 5. Ice Ages

Bolles, E. B. 1999. *Ice Finders: How a Poet, a Professor, and a Politician Discovered the Ice Age*. New York: Counterpoint.

Dawson, A. G. 1992. *Ice Age Earth: Late Quaternary Geology and Climate*. London: Routledge. 293 pp.

Hallam, A. 1989. *Great Geological Controversies*. 2nd edition. Oxford: Oxford University Press. 244 pp.

Imbrie, J. and K. P. Imbrie. 1979. *Ice Ages: Solving the Mystery*. Hillside, N.J.: Enslow Publishers. 224 pp.

C. J. Schneer, 1969. *Toward a History of Geology*. Cambridge, Mass.: M.I.T. Press.

Chapter 6. Origin of the Moon

French, B. M. 1977. *The Moon Book*. Baltimore, Md.: Penguin Books. 287 pp.

Hartmann, W. K., R. J. Phillips, and G. J. Taylor, eds. 1986. *Origin of the Moon*. Houston, Tex.: Lunar Planetary Institute. 781 pp.

Heiken, G. H., D. T. Vaniman, and B. M. French, eds. 1991. *Lunar Sourcebook: A User's Guide to the Moon*. Cambridge: Cambridge University Press. 736 pp.

Hockey, T. A. 1986. *The Book of the Moon: A Lunar Introduction to Astronomy, Geology, Space Physics and Space Travel*. New York: Prentice Hall. 273 pp.

Spudis, P. D. 1996. *The Once and Future Moon*. Washington, D.C.: Smithsonian Institution Press. 308 pp.

Taylor, S. R. 1975. *Lunar Science: A Post-Apollo View*. New York: Pergamon Press. 372 pp.

———. 1982. *Planetary Science: A Lunar Perspective*. Houston, Tex.: Lunar Planetary Institute. 481 pp.

Chapter 7. Asteroids and Meteorites

Burke, J. G. 1986. *Cosmic Debris: Meteorites in History*. Berkeley, Calif.: University of California Press. 445 pp.

Cunningham, C. J. 1988. *Introduction to Asteroids*. Richmond, Va.: Willmann-Bell. 208 pp.

Dodd, R. T. 1981. *Meteorites: A Petrologic-Chemical Synthesis*. Cambridge: Cambridge University Press. 368 pp.

Huggett, R. 1998. *Catastrophism: Asteroids, Comets and Other Dynamic Events in Earth History*. London: Verso Books. 262 pp.

Kowal, C. T. 1988. *Asteroids: Their Nature and Utilization*, New York: Ellis Horwood. 152 pp.

McSween, H. Y. 1999. *Meteorites and Their Parent Planets*. 2nd edition. Cambridge: Cambridge University Press. 310 pp.

Norton, O. R. 1998. *Rocks from Space*. 2nd edition. Missoula, Mont.: Mountain Press. 447 pp.

Sumners, C., and C. Allen. 2000. *Cosmic Pinball: The Science of Comets, Meteors, and Asteroids*. New York: McGraw-Hill. 190 pp.

Taylor, S. R. 1992. *Solar System Evolution: A New Perspective*. Cambridge: Cambridge University Press. 307 pp.

Wasson, J. T. 1985. *Meteorites: Their Record of Early Solar-System History*. New York: W. H. Freeman. 267 pp.

Chapter 8. What Heated the Asteroids?

Hutchison, R. 1983. *The Search for Our Beginning*. London: British Museum (Natural History). 164 pp.

Kerridge, J. F., and M. S. Matthews, eds. 1988. *Meteorites and the Early Solar System*. Tucson: University of Arizona Press. 1,269 pp.

McSween, H. Y. 1999. *Meteorites and Their Parent Planets*. 2nd edition. Cambridge: Cambridge University Press. 310 pp.

Wasson, J. T. 1985. *Meteorites: Their Record of Early Solar-System History*. New York: Freeman. 267 pp.

Chapter 9. Mesosiderites: Biography of a Shocked and Melted Asteroid

Dodd, R. T. 1981. *Meteorites: A Petrologic-Chemical Synthesis*. Cambridge: Cambridge University Press. 368 pp.

McSween, H. Y. 1999. *Meteorites and Their Parent Planets*. 2nd edition. Cambridge: Cambridge University Press. 310 pp.

Norton, O. R. 1994. *Rocks from Space*. 2nd edition. Missoula, Mont.: Mountain Press. 449 pp.

Wasson, J. T. 1985. *Meteorites: Their Record of Early Solar-System History*. New York: W. H. Freeman. 267 pp.

Chapter 10. Meteor Crater

Hoyt, W. G. 1987. *Coon Mountain Controversies: Meteor Crater and the Development of Impact Theory*. Tucson: University of Arizona Press.

Ley, W. 1968. *The Meteorite Craters*. New York: Weybright and Talley. 135 pp.

Mark, K. 1987. *Meteorite Craters*. Tucson: University of Arizona Press. 288 pp.

Melosh, H. J. 1989. *Impact Cratering: A Geologic Process*. New York: Oxford University Press. 245 pp.

Nininger, H. H. 1956. *Arizona's Meteorite Crater: Past–Present–Future*. Denver, Colo.: American Meteorite Laboratory. 232 pp.

Poag, C. W. 1999. *Chesapeake Invader: Discovering America's Giant Meteorite Crater*. Princeton, N.J.: Princeton University Press. 183 pp.

Chapter 11. The Lunar Crater Controversy—A Brief Retelling

Mark, K. 1987. *Meteorite Craters*. Tucson: University of Arizona Press. 288 pp.

Melosh, H. J. 1989. *Impact Cratering: A Geologic Process*. New York: Oxford University Press. 245 pp.

Mutch, T. A. 1972. *Geology of the Moon: A Stratigraphic View*. Revised edition. Princeton, N.J.: Princeton University Press. 391 pp.

Short, N. M. 1975. *Planetary Geology*. Englewood Cliffs, N.J.: Prentice-Hall. 361 pp.

Chapter 12. Dinosaurs and the Cretaceous-Tertiary Extinction

Alvarez, W. 1997. *T. rex and the Crater of Doom*. Princeton, N.J.: Princeton University Press. 185 pp.

Frankel, C. 1999. *The End of the Dinosaurs: Chicxulub Crater and Mass Extinctions*. Cambridge: Cambridge University Press. 230 pp.

Powell, J. L. 1998. *Night Comes to the Cretaceous: Dinosaur Extinction and the Transformation of Modern Geology*. New York: W. H. Freeman. 325 pp.

Raup, D. M. 1991. *Extinction: Bad Genes or Bad Luck?* New York: W. W. Norton. 210 pp.

Wignall, P. B., and A. Hallam. 1997. *Mass Extinctions and Their Aftermath*. Oxford: Oxford University Press.

Chapter 13. Recent Impacts: Tunguska to Shoemaker-Levy 9

Crovisier, J., and T. Encrenaz. 2000. *Comet Science: The Study of Remnants from the Birth of the Solar System*. Cambridge: Cambridge University Press. 173 pp.

Gallant, R. 1996. Sikhote-Alin revisited. *Meteorite!* 2, no. 1, 8–11.

Krinov, E. L. 1966. *Giant Meteorites*. Oxford: Pergamon Press. 397 pp.

Levy, D. H. 1995. *Impact Jupiter: The Crash of Comet Shoemaker-Levy 9*. New York: Plenum Press. 290 pp.

Mark, K. 1987. *Meteorite Craters*. Tucson: University of Arizona Press, 288 pp.

Spencer, J. R., and J. Mitton, eds. 1995. *The Great Comet Crash: The Impact of Comet Shoemaker-Levy 9 on Jupiter*. Cambridge: Cambridge University Press. 118 pp.

Sumners, C., and C. Allen. 2000. *Cosmic Pinball: The Science of Comets, Meteors, and Asteroids*. New York: McGraw-Hill. 190 pp.

Chapter 14. Tektites: A Glass Menagerie

Barnes, V. E., and M. A. Barnes, eds. 1973. *Tektites*. Stroudsburg, Pa.: Dowden, Hutchinson & Ross. 445 pp.

Dodd, R. T. 1986. *Thunderstones and Shooting Stars: The Meaning of Meteorites*. Cambridge, Mass.: Harvard University Press. 196 pp.

O'Keefe, J. A. 1976. *Tektites and Their Origin*. Amsterdam: Elsevier Scientific Publishing Co. 254 pp.

———. ed. 1963. *Tektites*, Chicago: University of Chicago Press. 228 pp.

Chapter 15. Rings and Shepherds

Beatty J. K., C. C. Petersen, and A. Chaikin, eds. 1998. *The New Solar System*. 4th edition. Cambridge: Cambridge University Press. 432 pp.

Impey, C., and W. K. Hartmann. 2000. *The Universe Revealed*. Pacific Grove, Calif.: Brooks/Cole. 636 pp.

Jones, B. W. 1999. *Discovering the Solar System*. Chichester: Wiley & Sons. 416 pp.

Littmann, M. 1990. *Planets Beyond: Discovering the Outer Solar System*. New York: John Wiley & Sons. 319 pp.

Seeds, M. A. 1999. *Foundations of Astronomy*. Belmont, Calif.: Wadsworth Publishing Co. 640 pp.

Taylor, S. R. 1992. *Solar System Evolution: A New Perspective*. Cambridge: Cambridge University Press. 307 pp.

Chapter 16. The Search for Life on Mars

Davies, P. 1999. *The Fifth Miracle: The Search for the Origin and Meaning of Life*. New York: Simon and Schuster. 304 pp.

Dick, S. J. 1996. *The Biological Universe: The Twentieth-Century Extraterrestrial Life Debate and the Limits of Science.* Cambridge: Cambridge University Press. 578 pp.

———. 1998. *Life on Other Worlds: The 20th-Century Extraterrestrial Life Debate.* Cambridge: Cambridge University Press. 290 pp.

Drake, F., and D. Sobel. 1992. *Is Anyone Out There? The Scientific Search for Extraterrestrial Intelligence.* New York: Delacorte Press. 272 pp.

Dyson, F. 1999 *Origins of Life.* 2nd edition. Cambridge: Cambridge University Press. 100 pp.

Goldsmith, D. 1997. *The Hunt for Life on Mars.* New York: Penguin Books. 267 pp.

Jakosky, B. 1998. *The Search for Life on Other Planets.* Cambridge: Cambridge University Press. 326 pp.

Koerner, D., and S. LeVay. 2000. *Here Be Dragons: The Scientific Quest for Extraterrestrial Life.* Oxford: Oxford University Press. 264 pp.

Sagan, C. 1980. *Cosmos.* New York: Random House. 365 pp.

Schopf, J. W. 1999. *Cradle of Life: The Discovery of Earth's Earliest Fossils.* Princeton, N.J.: Princeton University Press. 367 pp.

Shapiro, R. 1999. *Planetary Dreams: The Quest to Discover Life beyond Earth,* New York: John Wiley & Sons. 306 pp.

Sheehan, W. 1996. *The Planet Mars: A History of Observation and Discovery.* Tucson: University of Arizona Press. 270 pp.

Walter, M. 1999. *The Search for Life on Mars.* Cambridge, Mass.: Perseus Books. 170 pp.

Chapter 17. Panspermia

Aczel, A. D. 1998. *Probability 1: Why There Must Be Intelligent Life in the Universe.* New York: Harcourt Brace & Company. 230 pp.

Davies, P. 1999. *The Fifth Miracle: The Search for the Origin and Meaning of Life.* New York: Simon and Schuster. 304 pp.

Dick, S. J. 1996. *The Biological Universe: The Twentieth-Century Extraterrestrial Life Debate and the Limits of Science.* Cambridge: Cambridge University Press. 578 pp.

———. 1998. *Life on Other Worlds: The 20th-Century Extraterrestrial Life Debate.* Cambridge: Cambridge University Press. 290 pp.

Dyson, F. 1999. *Origins of Life.* 2nd edition. Cambridge: Cambridge University Press. 100 pp.

Fry, I. 2000. *The Emergence of Life on Earth: A Historical and Scientific Overview*. Piscataway, N.J.: Rutgers University Press. 256 pp.

Jakosky, B. 1998. *The Search for Life on Other Planets*. Cambridge: Cambridge University Press. 326 pp.

Koerner, D. W., and S. LeVay. 2000. *Here Be Dragons: The Scientific Quest for Extraterrestrial Life*. Oxford: Oxford University Press. 256 pp.

Kutter, G. S. 1987. *The Universe and Life: Origins and Evolution*. Boston: Jones and Bartlett Publishers. 592 pp.

Schopf, J. W. 1999. *Cradle of Life: The Discovery of Earth's Earliest Fossils*. Princeton, N.J.: Princeton University Press. 367 pp.

Shapiro, R. 1999. *Planetary Dreams: The Quest to Discover Life beyond Earth*. New York: John Wiley & Sons. 306 pp.

Shklovskii, I. S., and C. Sagan. 1966. *Intelligent Life in the Universe*. New York: Dell Publishing Co. 509 pp.

Chapter 18. Paucity of Aliens

Achenbach, J. 1999. *Captured by Aliens: The Search for Life and Truth in a Very Large Universe*. New York: Simon and Schuster. 415 pp.

Aczel, A. D. 1998. *Probability 1: Why There Must Be Intelligent Life in the Universe*. New York: Harcourt Brace & Company. 230 pp.

Boss, A. 1998. *Looking for Earths: The Race to Find New Solar Systems*. New York: John Wiley & Sons. 240 pp.

Broecker, W. 1985. *How to Build a Habitable Planet*. Palisades, N.Y.: Eldigio Press.

Davies, P. 1995. *Are We Alone? Philosophical Implications of the Discovery of Extraterrestrial Life*. New York: Basic Books. 160 pp.

———. 1999. *The Fifth Miracle: The Search for the Origin and Meaning of Life*. New York: Simon and Schuster. 304 pp.

Diamond, J. 1992. *The Third Chimpanzee: The Evolution and Future of the Human Animal*. New York: Harper Perennial. 407 pp.

Dick, S. J. 1996. *The Biological Universe: The Twentieth-Century Extraterrestrial Life Debate and the Limits of Science*. Cambridge: Cambridge University Press. 578 pp.

———. 1998. *Life on Other Worlds: The 20th-Century Extraterrestrial Life Debate*. Cambridge: Cambridge University Press. 290 pp.

Drake, F., and D. Sobel. 1992. *Is Anyone Out There? The Scientific*

Search for Extraterrestrial Intelligence. New York: Delacorte Press. 272 pp.

F. Dyson, 1999. *Origins of Life.* 2nd edition. Cambridge: Cambridge University Press. 100 pp.

Goldsmith, D. 1997. *Worlds Unnumbered: The Search for Extrasolar Planets.* Sausalito, Calif.: University Science Books. 237 pp.

Gould, S. J. 1989. *Wonderful Life: The Burgess Shale and the Nature of History.* New York: W. W. Norton and Co. 347 pp.

Koerner, D., and S. LeVay. 2000. *Here Be Dragons: The Scientific Quest for Extraterrestrial Life.* Oxford: Oxford University Press. 264 pp.

Schopf, J. W. 1999. *Cradle of Life: The Discovery of Earth's Earliest Fossils.* Princeton, N.J.: Princeton University Press. 367 pp.

Shklovskii, I. S., and C. Sagan. 1966. *Intelligent Life in the Universe.* New York: Dell Publishing Co. 509 pp.

Simpson, G. G. 1964. The nonprevalence of humanoids. *Science* 143, 769–75.

Ward, P. D., and D. Brownlee. 2000. *Rare Earth: Why Complex Life is Uncommon in the Universe.* New York: Copernicus, an imprint of Springer-Verlag.

Zuckerman, B., and M. H. Hart, eds. 1995. *Extraterrestrials: Where Are They?* 2nd edition. Cambridge: Cambridge University Press. 239 pp.

Chapter 19. Human Response to First Contact

Achenbach, J. 1999. *Captured by Aliens: The Search for Life and Truth in a Very Large Universe.* New York: Simon and Schuster. 415 pp.

Christian, J. L., ed. 1976. *Extraterrestrial Intelligence: The First Encounter.* Buffalo, N.Y.: Prometheus Books. 303 pp.

Dick, S. J. 1996. *The Biological Universe: The Twentieth-Century Extraterrestrial Life Debate and the Limits of Science.* Cambridge: Cambridge University Press. 578 pp.

———. 1998. *Life on Other Worlds: The 20th-Century Extraterrestrial Life Debate.* Cambridge: Cambridge University Press. 290 pp.

Harrison, A. A. 1997. *After Contact: The Human Response to Extraterrestrial Life.* New York: Plenum Press. 363 pp.

Jones, B. W. 1999. *Discovering the Solar System.* Chichester: John Wiley & Sons. 416 pp.

McSween, H. Y. 1999. *Meteorites and Their Parent Planets*. 2nd edition. Cambridge: Cambridge University Press. 310 pp.

Tough, A., ed. 2000. *When SETI Succeeds: The Impact of High-Information Contact*. Washington, D.C.: Foundation for the Future.

Wasson, J. T. 1985. *Meteorites: Their Record of Early Solar-System History*. New York: W. H. Freeman. 267 pp.

INDEX

○

pallasites, 119
Pangaea, 79
panspermia, 260
parallax, 18
Pasteur, Louis, 261, 275
pearls on a string, 190
Penfield, Glen, 175
Peregrinus, Petrus, 48
periodic comet showers, 178
period-luminosity relationship, 25
permafrost, 247
Permian-Triassic boundary, 178
Perraudin, Jean-Pierre, 66
Phobos, 237
Piazzi, Guiseppe, 101
Pilgrim, Ludwig, 77
Pioneer 10, 227, 317, 323
Pioneer 11, 221, 227, 317
Pioneer plaque, 317–19
Pitakpaivan, Kaset, 212
planetary magnetic fields, 11
planetary nebulae, 5
planetary rings, 13, 217; origin of, 232
planetesimals, 6, 7
plankton, 82
plate tectonics, 58, 289
plesiosaurs, 168, 169
Plot, Robert, 167
Pluto, 13, 33
plutonium fission tracks, 109
polycyclic aromatic hydrocarbons (PAHs), 252
precession, 75
Project Ozma, 301
prokaryotes, 291
proton precession magnetometer, 57
Prutenic Tables, 20
pterosaurs, 168, 169
Ptolemy, 19
pulsars, 309
pyrolytic release experiment, 248

radiation pressure, 262
radioactive decay, 35, 37

radionuclides, long-lived, 122
Raëlian movement, 268
rare-earth elements, 134
rarefaction waves, 157
Raup, David, 178
Redi, Francesco, 261
refractory inclusions, 7, 133
resonances, 232
Ridpath, Ian, 312
Ries Crater, 211
ringlets, 221
ring particles, 224
rings, planetary, 13, 217
Ringwood, Ted, 90
Ritchey, George, 26
Roche, Édouard, 221
Roche limit, 190, 221
Roy, Sharat, 275
RR Lyrae stars, 26
rubble piles, 127
Rudolphine Tables, 22
runaway greenhouse effect, 286
Ruskol, Elena, 92
Russell, Henry Norris, 239, 242
Ryle, Martin, 312, 319

Sagan, Carl, 249, 253, 262, 311, 317
satellites, shepherding, 223, 232
Saturn, 216; Cassini Division in rings of, 217, 221; Encke Division in rings of, 218, 221; rings of, 217–21
Saturn, moons of: Dione, 220; Enceladus, 220, 224; Mimas, 219; Pandora, 223; Prometheus, 223; Titan, 216
Scheuchzer, Johann Jacob, 64
Schiaparelli, Giovanni, 237
Schimper, Karl, 68
Scotti, Jim, 189
secular variation, 50
Sepkoski, Jack, 178
SETI, 306
shale balls, 151
Shapley, Harlow, 25